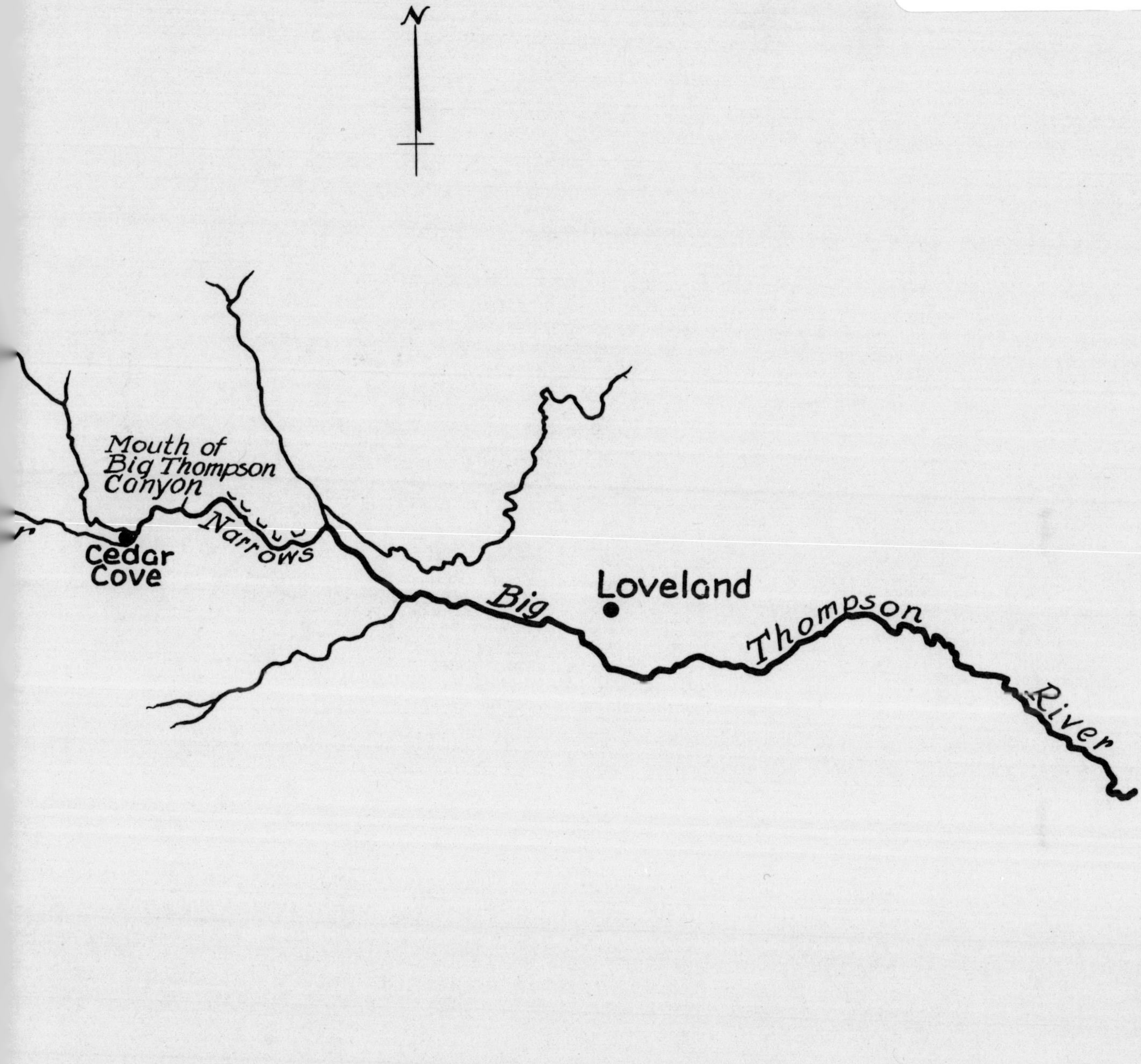
N
Mouth of
Big Thompson
Canyon
Cedar
Cove
Narrows
Loveland
Big
Thompson
River
5
10
15 MI.
5
10
15 KM

BIG THOMPSON:

For Susan

BIG THOMPSON:
Profile of a Natural Disaster

By David McComb

PRUETT PUBLISHING COMPANY
Boulder, Colorado

Library of Congress Cataloging in Publication Data

McComb, David G.
Big Thompson.
Includes bibliographical references.
1. Floods—Colorado—Big Thompson Canyon. 2. Big Thompson Canyon—History. 3. Big Thompson River—Flood, 1976. I. Title.
F782.B45M33 987.8'68 80-15397
ISBN 0-87108-558-5

First Edition

1 2 3 4 5 6 7 8 9

Printed in the United States of America

All photographs not otherwise credited were taken by the author.

Contents

The Big Thompson cloud. Mark L. Griswold took this rare photograph in the early evening just before the flood from the Rawah Wilderness, northwest of Fort Collins. He was working on maps for the United States Geological Survey and for his M.S. degree from Colorado State University. The cloud appeared distinctive and yet innocuous to him. About two months later he realized that he had photographed the thunderhead which caused the flash flood. *Photograph courtesy of Mark L. Griswold.*

Introduction

This study of the Big Thompson disaster began nearly two weeks after the rampage in the canyon. It would not have started even then except for my twelve-year-old, brown-eyed daughter, Susan. Together we watched as reporters unfolded the tragedy only a few miles away. As I left for my office one morning, Susan stopped me and asked, "Why don't you do some work on the Big Thompson?" "Well," I replied, "because I am an urban historian, interested in disasters that strike cities." "Yes," she said, with the usual adolescent obstinance, "but you are also an oral historian. This disaster is in your backyard, and you ought to do something." Kids are full of 'you ought's,' so I shrugged and went on my way.

As the day wore on, however, Susan's words took effect. She was right. I should do something. I called William Marshall, the executive director of the State Historical Society to request endorsement for a Big Thompson disaster oral history project. "Great!" he said, after I outlined the tape-recorded interviews I had in mind. "Why didn't anyone think of it before?" He gave his blessing. George Dennison, who was just beginning his career as a Colorado State University administrator, and President A. R. Chamberlain, also saw the worth of the project. They released me from fall semester classes, gave me some expense money, and told me to go ahead.

Fortunately, I had expert advice and help. Colorado possessed, oddly enough, several of the world's outstanding sociologists of disaster—Thomas E. Drabek of Denver University; J. Eugene Haas, then of the University of Colorado; and John R. Brouillette of Colorado State University. Along with Lee Scamehorn, an historian at the University of Colorado, they provided general information about natural calamities and constructive criticism on the project. The Experiment Station at C.S.U. provided money for editing and transcription of tapes. Mary Alice McComb edited the interviews, checked the book manuscript for errors, constructed a map of the canyon, and demonstrated the usefulness of an additional camera at the disaster site. The result—forty-one oral history interviews with victims or people involved in the rescue/recovery effort and 250 photographs. The Colorado Historical Society, which is the repository for the interviews, now possesses information which would not have existed otherwise. The tapes are the basis for this book. Colleen T. Kain, a talented and important friend, generously took the time to read and criticize the manuscript. Many people thus shared their experience and future Coloradoans, if they heed the lessons of the Big Thompson, will benefit. The project and this book, however, were inspired by a bright twelve-year-old girl. It was Susan's idea.

At the turn of the century a rough wagon road wound through the Big Thompson Canyon. *Photograph courtesy of Western History Department, Denver Public Library.*

Chapter I

Preparation: By Time and Degree

The Big Thompson Canyon had been well scouted by the time Mariana Modina, a squawman from Taos, arrived in 1858 to establish a toll bridge and general store for the stage line. He was the first settler in the valley. Earlier, Ute Indians, seeking access to the plains, had marked a trail through the canyon. In 1840, Kit Carson followed their path to a winter camp in Estes Park. In 1810, David Thompson, an engineer, astronomer, and agent for the Northwest Fur Company, explored the area.[1] The Englishman left his name on the land, but the canyon with its awesome beauty was far older than these recent human intrusions.

The Rocky Mountains appeared sixty to seventy million years before when the cracking and buckling of the earth's crust sent layers of metamorphic and sedimentary rock skyward. Over time, through the eroding run-off of rain and snow and periodic catastrophic cloudbursts which scourged the rivercourse with tons of rushing water and rocks, the Big Thompson River etched its canyon through a natural seam in the mountains. It was not an unusual occurrence, and the same action took place elsewhere along the front range less spectacularly.[2] The climate provided moderately cold winters and warm summers with over 300 days of sunshine. The air was dry, however, with only fourteen inches of yearly rainfall.[3] Nature had established her pattern, and it was up to humanity to live in harmony with the rhythm.

To a traveler, particularly in a fast-moving automobile, the front range rises abruptly from the flatness of the plains with little warning. The mountains are simply and suddenly there as they have always been—steep, tempting, and unforgiving. They are as intriguing now as they were a hundred years ago. A few miles west of the mid-sized town of Loveland in north-central Colorado and across two low, red shale hogback ridges, the V-shaped mouth of the Big Thompson Canyon opens like a narrow crack in the granite barrier of the mountains.

The highway, giving the motorist no choice, plunges into the bottom of the wedge, following the twists and turns of the river. For two miles the stratified brown and rust colored walls stretch 500 feet perpendicularly from the edge of the road. To avoid an accident, a driver must concentrate on the center stripe, the hairpin turns, and the traffic. At such close range the stone walls fill the entire windshield and only by leaning forward and twisting can a person see the rim of the so-called Narrows.

It is easy to become anxious and nervous here. Human beings are dwarfed at the bottom of this chasm. It is claustrophobic; there is no quick escape. You are in a partially closed stone hand. What if, by a geological twitch, the hand should close completely catching you in its grasp? Earthquakes are not unknown. The rock of the walls is loose and rotten. Every year mountain climbers practicing their art slip in a scurry of stone to

The Narrows in the 1920s. *Photograph courtesy of Western History Department, Denver Public Library.*

Between 1902 and 1904 Loveland businessmen financed a one-lane dirt road through the canyon. *Photograph courtesy of Colorado Historical Society.*

injury or death. The Narrows are not to be trusted.

No one can settle or build a cabin in this region. It is too steep and barren. Even mountain shrubs have a hard time. But the road and the siphon provide evidence of human penetration and technical dominance. They are reassuring. The siphon spans the opening of the Narrows thirty feet over the highway. It is a straight steel pipe, nine feet in diameter and weighing 225,000 pounds. Originally painted silver, it emerges and enters the solid rock walls like a huge straight pin holding the mountains together.

The siphon, built in the 1930s, transfers water from the western slope of the Rockies to the eastern edge. It is part of the Charles Hansen feeder canal which supplies over 500 cubic feet of water per second to Horsetooth Reservoir located west of Fort Collins. The reservoir, canal, and siphon are all part of the Bureau of Reclamation's Colorado-Big Thompson Project which provides irrigation and urban water for northern Colorado. The Bureau monitors snowmelt and rainfall, collects the runoff, and pipes it across the Continental Divide through a series of lakes, dams, and power plants. The project is one of the great engineering accomplishments of the West; it allows concentrated settlement in an area of scarce water.[4]

Beneath the pylons of the siphon trickles the Big Thompson River. Except in Spring, water runs only a few feet deep and fifteen to twenty feet wide. The river happily rushes over rocks and gurgles through pools as it descends 8,000 feet from the high range to the South Platte on the prairie. The water is clear, numbing cold, and playfully sparkled by the sun. It has always been a superb trout stream—too good, in fact. The state has to maintain the fish population with 18,000 pounds of rainbow trout each year from the fish hatchery at Drake, midway up the canyon.[5] This is part of the cost of maintaining Colorado as a tourist attraction. Too many people like to fish and need to be encouraged by catching something once in a while.

The highway paralleling the river is also a tourist attraction. In 1902, Loveland merchants financed a $24,000 project to construct a one-lane, dirt road through the canyon. Completed in 1904, it widened the dusty cattle and Indian

The road of the upper canyon before pavement. *Photograph courtesy of Colorado Historical Society.*

The improvement of the highway marked the beginning of the modern use of the canyon. Tourists and businesses followed the highway. *Photograph courtesy of Colorado Historical Society.*

path and improved the passage with corduroy bridges. These low-level structures with logs laid crosswise to the road provided a semi-durable, if jolting, crossing of the river. The horses hated them and had to be led over. It was difficult for wagons to pass and the trip to Estes Park took two days with a night spent at the Forks Hotel halfway through the canyon. In time, particularly in 1908 and 1925, the state improved, widened, and reinforced the road. Early stage coaches and steam cars used it to transport guests to the Stanley Hotel in Estes Park. Horns blown in warning resounded off the rock walls as cars wheeled around the curves. A few intrepid individuals built vacation cabins along the route. Colorado hailed itself as the 'Switzerland of America.' Long's Peak became 'The American Matterhorn,' and the Big Thompson was 'The Loveland Grand Canyon.'[6]

During the 1930s Governor Ed Johnson fulfilled his promise to improve the road in the canyon. In 1938 it became a national highway, U.S. 34, beginning the modern era for the Big Thompson.[7] Motels, restaurants, campsites, picnic areas, vacation houses, tourist shops, and some all-year homes followed the highway. A smooth two-lane blacktop with high bridges and embankments built to resist floods now ran through the Narrows. The travel time to Estes Park was cut to half a day and there was no need to linger in the canyon. People did stop, however, to stretch, enjoy the scenery, fish, spend the night, and listen to the music of mountain water.

Although the closeness of the Narrows forbids the construction of a cabin, or even a gas station, U.S. 34, after several twisting miles, swerves sharply into the Cedar Cove area. Here, the canyon widens and there is room on both sides for trees, grass, and houses. It really is a cove, of sorts. The Utes liked to spend the winter in the small valley hidden from enemies and protected from the howling, cold wind overhead. There were timber, fish, and deer. Cedar Cove was settled in the 1890s for small-scale logging and cattle-raising even before the road pushed through.[8] In time motels came such as the Mont-

The improved roadway became U.S. 34 in 1938. *Photograph courtesy of Western History Department, Denver Public Library.*

rose Inn and the Park Terrace, and the restaurants near the river such as the Covered Wagon and the Canyon Inn. The Cove is green, protective, and beautiful. The Indians were right; it was a good place to live.

Beyond Cedar Cove the highway climbs halfway up the side of the valley and runs past the location of the Loveland power plant and Viestenz-Smith Park. This site is a tribute to the determination of one man, Charles Viestenz. He was a member of the Loveland City Council and, in 1911, had a dream of cutting electricity costs for his town by placing a hydroelectric generator on the Big Thompson. The power suppliers and the other council members blocked the move, but Viestenz bought the site in secret and gained citizen approval at a general election in 1914. Construction began in 1915, but because of World War I and other delays, it was not completed for ten years. In 1925, electricity began to flow from a trio of generators and water wheels housed in a stout concrete and brick building. Loveland purchased the adjoining 400 acres for a grassy, riverbank park with picnic tables scattered among the trees. Even though only five percent of Loveland's electric power came from the plant in 1976, it inexpensively serviced the people in the lower canyon.[9] Charles Viestenz had won his point.

Halfway through the passage the North Fork loops down to join the Big Thompson. The North Fork is the main tributary within the canyon; Buckhorn Creek, the other major tributary, joins the Big Thompson outside the mountains near Loveland. At the "Y" configuration there is again room for settlement. This was the location of the Forks Hotel. In 1905 it became the site of Drake, a village named for the senator who gave it a post office.[10] With the modern highway, Drake became a stopping place for gasoline, snacks, and Indian jewelry. There were highway maintenance facilities, a state fish hatchery, and several hundred people living in small frame houses and cabins.

Among the earliest residents were D. T. and Lillian B. Pulliam. In 1909, they thought they would build a cabin on one of the flat, inviting meadows. It looked pleasant and easy, but they found something strange in the meadow and heeded its warning. On their spot was a large log which somehow had floated and lodged high on the land. They built instead on the hillside and their old cabin survived and endured.[11]

The highway divides at Drake. To the north, following the tributary, the road leads to the old village of Glen Haven. In recent years, Bill and Doris Wells turned the Glen Haven Inn into a nationally famous gourmet restaurant. It was a place to celebrate an anniversary or job promotion. You could spend the night at the Inn, buy groceries nearby, or browse at Calico Kate's gift shop. Hidden in the surrounding pine forests were some 300 summer homes.[12]

After Glen Haven the road climbs, switches sharply back and forth, and emerges at the side door of Estes Park. When Glen Haven was a logging camp, this Devil's Gulch approach was the main entrance to the mountain resort area.[13] The main route in 1976, however, followed the course of the Big Thompson River.

Above Drake U.S. 34 winds past the small settlements of Waltonia, Grandpa's Retreat, Glen Comfort, and Loveland Heights. Vacation cabins cluster beside the highway and motels crowd the river. There are a variety of access bridges, many of them obviously home-built. Between islands of habitation the sloping mountainsides are covered with forests of ponderosa pine and Douglas-fir. A few dead and rust-colored trees scattered through the forests mark the presence of the deadly pine bark beetle. There are some aspen along the route, but not many. Everywhere are grasses and shrubs of mountain mahogany, bitterbrush, chokecherry, rabbitbrush, and gooseberry. Wildlife includes Abert's squirrel, mule deer, black bear, coyote, marmot, weasel, golden eagle, skunk, horned owl, bobcat, chipmunk, and mountain bluebird.[14] These are hard to notice from a speeding automobile.

Rising quickly from the canyon and over a hill, U.S. 34 enters the Estes Park basin where Olympus Dam and Lake Estes form part of the Colorado-Big Thompson Project. The tourist town of Estes Park nestles in the basin. Above Estes Park lies Rocky Mountain National Park. Established in 1915, it serves 2,500,000 visitors a year and boasts of Trail Ridge Road, a spectacular drive over the Continental Divide. At the western end of Trail Ridge stands Milner Pass, the headwater of the Big Thompson River.

Altogether the river drains 830 square miles, three-quarters of that above the mouth of the canyon.[15] Within the canyon itself, in the summer of 1976, there were 2,500 to 3,000 people living or visiting. There were about 1,400 houses, but no condominiums as might be found at Vail or Aspen. On a normal summer day over 9,000 automobiles traveled U.S. 34 through the twenty-five-mile stretch of the canyon.[16]

By 1976 the Big Thompson was oversettled and unwisely used. People constructed homes and businesses in dangerous places. They were drawn by the natural, innocent beauty of the river. Through ignorance and short memory, people failed to read the warnings present on the land and in the records. They did not notice the log in the meadow and wonder how it got there. This set the scene for tragedy.

In a 1971 study, the Army Corps of Engineers stated that the Big Thompson flooded, on the average, every six years. There had been a stream gauge east of Drake operating since 1887; the Corps experts knew what they were talking about. They were concerned about Loveland, a population center, and commented, "A cloudburst in the lower basin could result in flooding at Loveland in two to three hours. A storm in the upper basin would produce high waters in Loveland in from 10 to 20 hours."[17]

There had been large floods in the past. Twelve of them were on record, in fact.[18] In 1864 overflow from the Big Thompson spread a mile wide at Loveland and left a legacy of mountain tree seeds to sprout on the plains.[19] Thirty years later, after a heavy downpour, water extended a "half-mile wide" and took out Loveland's water supply dam at the mouth of the canyon.[20] High water at Drake in 1906 washed away a bridge, and in 1919 a cloudburst took out more bridges and isolated the Forks Hotel.[21] Interestingly, there was a report of a ten-foot wall of water in 1919, a phenomenon also reported in 1976.

Lessons were learned in 1919. The Loveland *Daily Herald* stated, "Since then the road has been changed and all the low bridges elevated until now it is almost impossible for any such damage to take place."[22] People often underestimate nature, however, and heavy rain at night five days later pushed water twelve to fifteen inches over the road causing $200,000 in damages. There was a rumor that Lake Loveland was forty feet too high. It was only eight feet over and no threat. There was also fear of looting, so the local police department issued a public warning and dispatched three deputies to prevent it.[23]

The worst of the old floods came in August, 1951, when a four-inch rainfall at Cedar Cove broke the reservoir on Buckhorn Creek, seven miles west of Loveland. An eight-foot wall of water rushed over U.S. 34 sweeping away a Denver family traveling in their car. The flood killed seven people, left five families without homes, knocked out telephones, and deposited mud deep enough to cancel the annual county fair. Once again the flood caused $200,000 in damages and sightseers caused traffic jams on the highway.[24] You might think people would want to leave, but that was not the case. They wanted to see what had happened. The phenomena of sightseers, fear of looting, and rumor are all characteristic of modern disasters.

In the 1970s flash floods became nature's number one killer in the United States. The causes are inefficient warning systems, imprecise weather forecasting, and human fatalism.[25] It would seem illogical to build in an obvious floodway, but when no one there remembers a severe flood, there is no law to prevent it, and things look peaceful, why not? Living beside a bright, cheerful mountain trout stream is enticing.

Coloradoans, of course, knew about flash floods. In 1864 a "wall of water" rolled down Cherry Creek, killing eleven people and carrying away Denver's City Hall and all its records.[26] A hundred years later, in 1965, the South Platte jumped its banks causing $400,000,000 in damages to Denver. It could do it again, and incidentally, the cost would be greater. In 1904, a Missouri Pacific passenger train fell into the flooding Arkansas River when a bridge collapsed. Some eighty-eight to one hundred people drowned; it was the worst Colorado disaster—until 1976.[27]

People are not unintelligent. They do prepare for emergencies—somewhat—despite the cliche that a disaster must strike three times before society does something about it. Under the fright of nuclear attack in the 1950s and 1960s, Civil Defense provided a ghost organization with contingency plans of action. Officials stored water and food in public buildings, posted triangular designation signs and marked urban evacuation routes. In time, however, fear subsided, civil defense plans remained on dusty shelves, and cocktail conversation shifted away from hypothetical debates about repelling your stricken neighbor at the door of your private shelter during nuclear holocaust.

For Don Hataway, City Manager of Loveland, there was no time to consult the civil defense guide when the Big Thompson struck. Besides, the plans were designed for gradual flatland floods and for warfare. They did not fit this situation and he did not use them.[28] The part-time civil defense director in Larimer County did play an initial role, however, so earlier defense planning possessed some merit.

After the South Platte ravished almost half of the counties in Colorado in 1965, a shift of emphasis occurred at the state administrative level of civil defense. Before, there had been only slight consideration of natural disasters; afterwards, concern amounted to nearly half the effort. Logan Rappe, the state disaster coordinator, encouraged county planning and cooperation. He established a file of two hundred independent search and rescue teams. He learned how to aid counties when they overextended their capacity. He knew when and where to ask for help. Rappe thus established his office as an emergency contact point at the state level.[29] This was a correct step.

Emerging from this same widespread interest in flood protection and also from a sharpening concern about the quality of the environment, in 1973 the state legislature passed a land use bill and established a land use commission. A legislative study, in 1972, underscored considerable land development in eleven of the thirteen front range counties and pointed to the Big Thompson as an area of risk. Seventy-three others possessed higher priority and there was little money for study purposes.[30] The general consensus was that people should not reside within floodplains. It made sense. Laws existed to prevent it, but surveys had to be made, and hearings set up. Citizens had to have a chance to express opinions. It all takes time and money. Nonetheless, again, the land use legislation was a correct step.

The Corps of Engineers concluded in their 1971 study for Loveland, "To the many people who did not experience the 1951 flood, this report should create a respect for the awesome power of the flooding river."[31] Therefore, the potential danger of the Big Thompson was known and a major flood predictable. Yet, the best means of preventing catastrophe—keeping people out of the way—was not possible unless citizens were deprived of their property. It can be done, of course, through the power of the state, but no one likes to do it. Eminent domain conflicts with our cherished traditions of private property. In other cases, persons already established in a floodplain have been allowed to remain. It is a compromise. According to Larry Timm, Larimer County planner, "Since there wasn't much pressure for development and most of the development was there already, passing a regulation wouldn't have done much good." So, there was no hurry to zone the Big Thompson for flood protection. Timm added, however, "We knew it wasn't a real good situation."[32]

For unavoidable risk and unaffordable loss, the classic defense is insurance. The burden was too great for private companies, but the federal government had a program. Following the Rapid City flood of 1972 ideas about federal relief policies changed. To avoid direct handouts—an American political goal even in the days of the Great Depression—people were asked to indemnify themselves. The Federal Insurance Administration acted as insurer with policies available through authorized, private agents. The cost was reasonable: for a residence, $.25 per $100 valuation and $.35 per $100 valuation on the contents. The maximum was $35,000. For a business the rates were $.40 per $100 for the structure and $.75 per $100 for the contents. The maximum was $100,000.[33]

To be eligible, a community had to submit to a floodplain study and pass regulations to protect people and buildings in the flood zone. Because of danger in the Poudre River-La Porte area north of the Big Thompson all of Larimer County became eligible on July 2, 1974. Even though the La Porte segment was the only one studied in detail, people throughout the county could purchase insurance according to F.I.A. guidelines.[34] Only one person in the Big Thompson chose to do so, and he collected $6,000 for a flooded basement. Residents said they had no knowledge of the insurance; agents claimed otherwise. "It's very difficult to sell flood insurance to someone in the mountains who has a home on a slope," said one salesman. "They can't imagine how water could get to them."[35]

Captain John Englebert organized search and rescue teams before the disaster. One of the heroes of the disaster, he was described by Sheriff Robert Watson as "The Paul Bunyan of Larimer County." *Photograph courtesy of the* Triangle Review.

Sheriff Robert Watson directed the efforts of the county during the emergency rescue and evacuation. His authority was unquestioned. *Photograph courtesy of* Triangle Review.

The ultimate loss, however, is not property, but life. There had been some general preparation to rescue and protect people, even though no one imagined a disaster of such magnitude. Most significant was the Larimer County Sheriff's Office. Common problems for the office are forest fires and the rescue of lost hikers. Both require large numbers of trained volunteers. In 1969, under the direction of Captain John Englebert, the sheriff began a studied effort to expand and develop volunteer capability. There was little money for the work, but each year the volunteer program improved.

A motorized four-wheel patrol and a mounted posse existed when Englebert began and he built upon that. He created a standby group of 135 forest fire fighters and formed teams for mountain, underwater, and snowmobile rescue work. He encouraged emergency medical training, provided sessions on law enforcement, and providentially, one month before the flood, gave instruction on how to set up and operate a heliport. Volunteers provided not only their time, but also their equipment. Englebert could call them up and move them in fifteen minutes. He had a strong organization.[36]

Englebert was a tall man, a lean six feet five inches, who spoke with the cadence of a machine gun. He had the capacity to work himself to exhaustion, but it took him three days to do so. He wore blue jeans with his official shirt and his pickup truck was equipped with several radio networks for instant field communications. Sheriff Robert Watson said of his chief fire, search, and rescue man, "John, himself, has a degree in wildlife management and biology. And he has a fantastic knowledge of the terrain, the capabilities of a man, the endurance of an ox, the stride of Paul Bunyan. He's the Paul Bunyan of Larimer County, and I've always said so."[37] Such are the makings of a hero.

The Sheriff's headquarters are in Fort Collins, a twenty-minute drive north of Loveland and the Big Thompson. In Estes Park there is a substation and some very capable aid, when necessary, from the Rangers at Rocky Mountain National Park. The Rangers patrol 410 square miles of mountains. They are self-contained; that is, they carry and supply all of their own equipment. They are experienced and skilled. Federal law, moreover, permits the Rangers to work under local leadership in adjacent communities in the case of emergency.[38] A cordial relationship had grown between the Sheriff's Office and the park professionals. They were on call, available, and ready to help when needed.

A similar relationship had been worked out with an overlapping police authority, the Colorado State Patrol. Their jurisdiction is the highway, and their men provided the first accurate information about the flood. Two of their officers, moreover, had taken a two-week course in California which included a disaster exercise. This knowledge they shared with the Sheriff's Office.[39] Fort Collins and Loveland also had professional police and fire departments available. Even the Loveland fire volunteers had three men who completed a thirty-hour course in disaster rescue only days before it was needed.[40]

Behind these official groups stood several supporting agencies with strong reputations in disaster work, the Red Cross, Salvation Army, and more recently, the Seventh Day Adventists. These three organizations possess congressional authority to lend aid in the form of emergency food and shelter. In the past, the Red Cross and Salvation Army have tangled in their efforts to do good, squabbling over which one deserved primacy.[41] At the Big Thompson there was enough for all and the representatives on the spot simply split the tasks and cooperated to serve victims and rescuers.

Later, the Mennonite Disaster Service and the Interfaith Task Force stepped in to help people clean up and repair their property. These national organizations also had existence and purpose before the Big Thompson struck. In similar manner, the National Guard stood ready to help. Through proper authorization the Guard may be released for emergency service. Their policy is to respond quickly and to leave as soon as the job is over. They observe strict subservience to civil authority in order to avoid even a hint of martial law. Wariness of the military is in the American tradition. The National Guard, however, has access to unlimited numbers of people; it can respond rapidly and is self-contained.[42] In a sense, the Guard represents the ultimate emergency strength of the entire nation.

Perhaps the most important question in any plan to deal with catastrophe is: who is responsible? who is in charge? who has the authority? In other disasters there has been ambiguity of leadership. Some work was needlessly duplicated, while other critical tasks were neglected. Communications broke down and vital information was distorted by rumor and exaggeration. No one told the public what was occurring. Official agencies failed to coordinate. Emergent groups formed to meet needs, particularly for the job of search and rescue.[43] Basically, the history of other catastrophic natural events points to a frequent lack of leadership and control. There is an organizational disaster within the larger disaster caused by nature.

Under Colorado law, the county sheriff is in charge, and all other groups and officials from the governor on down take instructions from him. Obviously, if the sheriff is incompetent or incapacitated, or if the disaster cuts across jurisdictional lines, there may be difficulty. This was not the case in Larimer County. Leadership was clear, recognized, and competent.

All of the various groups—National Guard, Red Cross, Salvation Army, Seventh Day Adventist, Mennonite Disaster Service, Interfaith Task Force, Colorado State Patrol, National Park Service, Sheriff's Office, and others—therefore, possessed a certain level of readiness and direction. The organizations were prepared, in a broad sense, for emergencies and waited for specific incidents to exercise their capability. It would be impossible to prepare for all eventualities, but it was not as if the Big Thompson disaster occurred with no emergency plans in effect. Preparation was general, however, not specific.

At the mouth of the canyon, the Narrows returned to a primordial state. *Photograph courtesy of Fort Collins* Coloradoan.

Chapter II

Disaster: The Big Thompson Storm

Thunderstorms are normal for Colorado on summer afternoons. They are anticipated. Hikers and mountain climbers start early and avoid high points after the noon hour. The danger of being struck by lightning is too great. Moist air from the south sweeps upward along the front range forming thunderheads. Showers of short duration, intense and often with hail, occur as fast-moving winds aloft carry the storms eastward out over the plains. That is the usual pattern.[1]

Once in a while the high winds are weak or slow moving, allowing a storm to stall over an area. This is what happened on the early evening of Saturday, July 31, 1976. From the Gulf of Mexico, across Texas and from the Pacific over Arizona unusually moist air moved into Colorado against the front range. From Canada a polar front dropped diagonally across the northeast corner of the state and reversed the normal weather pattern. The moist air propelled at forty miles per hour by strong, southeasterly surface winds slammed into the foothills and boiled upward through the shallow frontal inversion. There was nothing to stop it and, like steam bursting through a ruptured pipe, the wet surface air mushroomed skyward to form a massive thunderhead towering 62,000 feet.

There were no high westerly winds to move it, and so it stood over the upper Big Thompson Canyon rumbling and freeing itself of its heavy burden of moisture. Between 6:30 and 10:30 p.m. twelve inches of rain fell on the canyon before the storm slowly drifted northward into Wyoming. This almost equalled the total annual rainfall. Observers on the plains at dusk noticed the dark, twelve-mile long system, the excessive amount of lightning, and a premature sunset at 7:00 p.m.[2]

It was startlingly similar to the 1972 Rapid City storm. Both occurred on an early summer evening, hit a tourist area, and had the same meteorological features. At Rapid City the storm broke a dam, killed 237 people, injured 2,932, and caused $100,000,000 in damages.[3] Nature was repeating herself in a different location.

The water fell heavily, in immense quantity, punctuated by deafening booms of thunder and blinding flashes of lightning. Captain William Thomas of the Colorado State Patrol who was at the mouth of the canyon and standing on the highway described the downpour:

> This was the darndest rain I was ever in. Raindrops as big as the end of the average man's thumb, coming straight down, lukewarm, and not an ounce of wind. And it caused a spray on the pavement, like, you know, they'd bounce, splatter about four feet high. It looked like you were just wading in water, and really, it was nothing but this heavy spray. It was hard to breathe, incidentally, there was so much water in the air.[4]

In the mountains there was little to hold it back. The steep rocky hillsides were barren or covered with thin, easily washed soil. The rain hit and quickly moved into established rivulets and

The complete disappearance of U.S. 34 through the Narrows was awesome to observers. *Photograph courtesy of Loveland* Reporter-Herald.

washes. Water flowing in broad sheets across small meadows left the grass beaten down as if pressed under a giant's foot. As it rushed downhill the runoff picked up soil, small rocks, twigs, leaves, and pine needles. The increasing force of the water uncovered and dislodged larger stones tumbling them along abrasively. Small trees and then larger ones on the slopes slipped away as the soil at the roots disappeared. The debris jammed in the gullies and wedged against boulders. It backed up water like a beaver dam.

Then, under increased and continued pressure, the wedge would give way with a surge of water and rolling branches. This wave had the power to break bridges. From a thousand washes, gullies, and rivulets the rain coursed into the larger arteries of the system—Fox Creek, Cedar Creek, Devil's Gulch—until it reached the North Fork and the Big Thompson.

Here, the water, heavily freighted with sand and debris, moved with the force of liquid concrete. Boulders ten feet in diameter ground and struck together underwater with a hollow "thock." Trees two feet thick rushed along, wedged, clung, tore at the bank, and moved on. U.S. 34, subjected to landslides from above and undermined from below, fell apart in countless places. As the flood moved downstream, grinding, tearing, ripping, it slowed around bends to gouge away the outside curve. The river boomed and rumbled.

Don McKee had a protected cabin near the bank which became a refuge for thirty people. He said of the sound: "The noise of the water was deafening. The whole mountain seemed to tremble and you were reminded of the noise of jet engines being tested. After awhile the continuous roaring beats against and into you, and makes you sick at the stomach. It disorients you."[5]

The thick, cocoa-colored water carried houses, cars, hissing propane tanks, mobile homes, and travel trailers on its surface, but not for long. The surface was not smooth and untroubled. Churning and plunging, it boiled over rolling boulders. The river pulled floating cars underneath. It twisted, scraped, pounded, and ripped them to pieces in a massive grinder of rock, water, and debris from which there was no escape. Sheriff Watson observed, "When bulldozers were trying to dredge landfill material from the streambed later, they found the remains of a small car four feet under the sand and silt and rocks."[6]

It was pitch black during the storm, but frequent horizontal lightning cracked through the rain-drenched canyon with strobe-light brilliance. It added to the cacophony of noise resounding between the mountain walls. There was thunder in the sky as well as in the river. It was a night of immeasurable terror for living things.

The water in the Big Thompson began to rise at 7:00 p.m. and peak stages struck Glen Comfort at 8:00, Waltonia at 8:30, Drake at 9:00, Cedar Cove at 9:30, and the mouth of the canyon at 11:00. At Glen Haven two peaks occurred, at 7:30 and 9:00 p.m. Thus, the North Fork surge did not hit Drake at the same time as the Big Thompson. This lessened the blow at the confluence, but sustained the downstream peak for a longer period. It lasted until after midnight at the mouth, seven miles distant.[7]

The water level rose nineteen feet above the normal depth of eighteen inches and poured at an incredible 31,200 cubic feet per second from the mouth of the canyon. Normal flow is 165 cubic feet per second.[8] The river moved twenty feet per second with unbelievable power. Silt-laden water carries more force than clear water because it has greater density. It took out the siphon between 10:15 and 10:30 p.m. when a floating object, possibly a house trailer, knocked down the supports. With the water inside, the siphon weighed 989,000 pounds. The Big Thompson carried it a quarter mile downstream and smashed it into the end of a house.[9] That was power.

On its rampage, the river destroyed 418 homes and 52 businesses. It harmed 138 other homes and caused $35,500,000 in total damages. Half of this amount involved the destruction of U.S. 34 and other roads. It killed 145 people, six of whom are still on the missing list. Forty-one of the victims lived in the canyon; the rest were outsiders. Their ages ranged considerably, but thirty-six percent were over fifty years.[10] It was the worst natural disaster in Colorado history.

There might have been better warning. At the time two groups had responsibility and authority to alert people about impending danger—the National Weather Service and the police. The N.W.S. failed; the police succeeded to the limit

U.S. 34 ended abruptly at the opening to the Narrows. This view, taken from the location of the siphon, looks eastward toward Loveland. The siphon lies downstream on the north bank. *Photograph courtesy of Fort Collins* Coloradoan.

The force of the water carried the siphon 600 yards downstream into the side of a house. *Photograph courtesy of Fort Collins* Coloradoan.

of their experience and saved, probably, several hundred lives.

The forecasting of weather, as everyone knows and jokes about, is an imprecise business. Yet, the meteorological conditions that spawn tornadoes, hurricanes and flash floods are well known. The extreme local nature and quickness of flash floods, supposedly, prevent effective prediction. That is the excuse. Tornadoes, however, are the same way. In 1950, the N.W.S. did not issue tornado warnings. Now they do, and in the twenty-seven tornado-prone states the level of warning, forecasting, and preparation is high, according to the service.[11]

There is a dependency upon local observers to confirm the existence of a flash flood. The Weather Service had set up a network of observers in Colorado, but they had neglected to maintain it, according to a National Oceanic and Aeronautics Administration investigative report. The annual visit for training and recruitment of observers in 1975 did not occur due to lack of funds. There were, however, unused 1976 monies. Potental sources of reliable and available information on rainfall, such as the Bureau of Reclamation, had not been utilized. As a result, not a single eyewitness confirmation of the Big Thompson flood reached the Denver weather headquarters in time to be of aid.[12]

To make matters worse, the video transmitter in the remote radar station at Limon, Colorado had broken down the day before. It had malfunctioned eight times in the preceding eighteen months and weather information had to be relayed verbally by telephone to Denver. At the very least, as admitted by the Weather Service, the breakdown cost some precision in pinpointing the storm.[13] At 7:35 p.m. the N.W.S. warned of severe thundershowers with a possibility of flooding in eastern Larimer County. This was followed at 9:00 p.m. with an "alert" of possible local flooding. The first actual flash flood warning came at 11:00 p.m.[14] The alarm was too late, and the N.W.S. lamely said it was meant to warn people downstream.[15] The failure of their warning system was apparent.

Only the National Weather Service can issue a flash flood warning, but a local weather company in Fort Collins recognized the danger and went as far as the law permitted. John Henz, Vince Sheetz, and Michael Fritsch started a private meteorological consulting service in 1973 called Geophysical R. and D. It is one of about fifty such firms in the United States. Fort Collins Weather Service, a part of the company, provides local weather information to radio stations based upon National Weather Service data transmitted from Suitland, Maryland.

At noon on the day before the flood, Friday, the weather company cautioned over station KCOL of Fort Collins, "Unstable air in the high country can lead to flash flooding and plenty of cloud-to-ground lightning. So, be careful."[16] Henz and Scheetz discussed the close similarity between conditions they were observing over the Big Thompson and what had happened at Rapid City four years earlier. Concerned, they spent part of Saturday monitoring the incoming weather maps. They repeated their earlier warning about unstable air and at 5:30 p.m. Scheetz broadcast:

> Those of you planning to be outside this evening, you want to be sure to stay out of the washes, creeks, and other low-lying areas if they are in the vicinity of these large thunderstorms. At this time the biggest cells are located in north central Larimer County, south Jefferson County, and Prowers County south of Lamar. However, conditions are unstable enough for more storms to develop during the evening hours after this broadcast is over and you should monitor your radio for any new developments that might occur.[17]

Scheetz also stated in his commentary, "The slow movement of these isolated cells makes them potentially dangerous due to heavy rains being confined to a small area. The potential for flash flooding, even along small washes and dry creek beds, could be considerable."[18] These warnings were repeated at 8:59 and 9:27 p.m. At 9:37 p.m. Henz made a special statement about heavy rain in Estes Park and about flooding on the Poudre River north of the Big Thompson.[19] What they feared began to occur. The weathermen had done what they could. What these broadcasts achieved cannot be calculated. How many people listened? How many responded to what they heard? More is known about the accomplishment of the police.

The police are usually the key link in warning systems.[20] They have the task of determining the severity of the emergency, telling people of danger, and mounting a rescue operation. That is their job; it is their responsibility. At the Big Thompson, it took time to realize the enormity of the rainfall. Some people would argue that it took the police too long. On the other hand, unlike the National Weather Service, the police had little direct experience of this kind and had to adapt their training and learning to deal with the problem.

Sometime between 7:30 and 8:15 p.m. Saturday night—the time is not exact—Sergeant Pat McCosh, dispatcher at the Larimer County Sheriff's Office, received a telephone call from a woman reporting rocks on the highway at Waltonia. This was probably Virginia McGee who notified the dispatcher from her store at Drake. Such reports were common, routine, and McCosh did not bother to log the conversation. At the same time, the Colorado State Patrol, with headquarters at Greeley, got another citizen's call about a washout above Waltonia at Grandpa's Retreat. The Patrol Office directed Officer William Miller, who was cruising U.S. 36 south of Estes Park, to investigate. It also requested Officer Tim Littlejohn who was on the southern edge of Fort Collins to assist Miller. As the two cars entered the canyon from each end, the dispatcher, Jay Lorance, informed their supervisor, Sergeant Hugh Purdy, of the whereabouts of his men.[21]

Purdy and his wife, Isa, were watching the Olympic Games on television, but when he heard of the potential rock slides from the dispatcher, he replied, "Yeah, I'll be right up." As a superior officer he thought it his duty to follow his men into the canyon. He slipped a slicker over his civilian clothes, climbed into his squad car, and left his Loveland home.[22] Miller in the upper canyon, meanwhile, was running into trouble.

He reported overflow and asked for help to get people out. Littlejohn reached Drake, but Purdy told him to stop, set up a roadblock, and turn travelers back down the highway. As he carried out the order he heard Miller's voice over the radio. "Advise, we have a flood. The whole mountainside is gone. We have people trapped on the other side. I'm going to have to move out. I'm up to my doors in water! Advise, we can't get to them. I'm going to get out of here before I drown!" Miller abandoned his car and scrambled up the mountain to safety.[23] This was the first

This concrete bridge in the Narrows withstood the direct assault of the current.

real identification of the flash flood.

Events now began to move quickly for Littlejohn at Drake. The chunky, round-faced, youthful officer turned on his flashing lights, switched on his loud speaker, and began telling people to flee for their lives. They wanted to know more; they did not understand; they hesitated. The house lights blinked off and in the darkness as the river surged over the banks Littlejohn heard some children crying about water in their home. People began to leave. There was no time for fright and the patrolman eased his car up the grade of the road to Glen Haven. The river crept up, slacked off, rose again. The officer kept his wheels ahead of the water as it covered the roadway. He could hear the deep rumble of large boulders as they ground together in the dark water, the clatter of rocks bounding off cliff sides, and the splintering of wooden houses. Over his radio, he briefly talked to his Sergeant.[24]

Purdy listened to his officers and requested further information as he doggedly drove toward Drake. He was a determined and sometimes stubborn man, a twenty-six year veteran of the force. He wanted to know about the amount of water and where it was. No one knew. He drove on, but ordered the patrolman trailing him to turn back and stand by at the Narrows to cut off entry to the canyon. At 9:00 p.m. he reported a sudden rise in the river and told the Patrol to warn the people of Loveland and those east of the Narrows. Fifteen minutes later the dispatcher in Greeley recorded Purdy's final report. "I'm stuck. I'm right in the middle of it. I can't get out. About a half-mile east of Drake on the highway. Tell them to get out of that low area down below. And as soon as the water starts picking up . . . (static) . . . high ground . . . ." Silence. The dispatcher tried to recontact him without success. Later, Sergeant Purdy's body was found, downstream; his car remained where it was, crushed under a slide of mud and rock.[25]

Meanwhile, the Larimer County Sheriff's Office continued to receive calls. A former state patrolman told dispatcher McCosh that water was crossing the road above Drake; a special deputy reported problems at Grandpa's Retreat; and the Estes Park police requested help. McCosh alerted Sheriff Watson and asked Captain Terry Urista, who lived in Loveland, to investigate.[26] Clad in blue jeans, boots, and an officer's shirt, Urista, an ex-marine, unwittingly followed Purdy's route up the canyon road. The captain noticed the debris in the choppy river. It made him uneasy; it did not look right. He paused to caution some people in Cedar Cove and drove on up U.S. 34 where the road climbed the mountainside. He moved past the Loveland power plant, and as he rounded the curve, suddenly, his headlights reached into black nothingness. He slammed to a stop; the road was gone. It abruptly ended in a sheer drop. On the outside of a curve, the river had washed the road away.[27]

Urista retreated to Cedar Cove, turned on his loudspeaker, and tried to warn people above the roar of the river. He later recalled:

> I saw some headlights over across the river, and they have a little wooden bridge that goes across the river there by Cedar Cove, and to those homes. The road runs behind all those homes, and the homes are right on the river.
>
> And I saw some headlights over there, so I figured I'd run over in there and get on my P.A. and warn the people. And I pulled onto this little wooden bridge they've got, got my wheels right up to it, and I took a look at it. The river was right up to the bottom of it.
>
> There was still a lot of junk in there and everything, and I had some second thoughts about it.

I pulled it into low gear and I was just going to shoot across the bridge. I figured I'd be okay; I'd be cool.

About the time I got ready to put my foot into it, put it into low gear and just ready to cram it to the floor, the upriver side of that bridge started to sink. And a little water started lapping over, and then a pine tree about two feet in diameter and about forty-foot long came through and just wiped that bridge out right in front of me. . . .

So, I put it in reverse and got back up on the roadway, and got on the P.A. and yelled in the P.A. for the people on the other side of the river to get out, get to high ground.[28]

He had no way of knowing if his voice was heard. It was. Afterwards he joined a fellow officer, James Garcia, also there to help whomever he could. Urista told his office to evacuate the low areas of Loveland. His own home was in danger, but he was marooned. An A-frame house had floated and crashed on top of the bridge leading from the Narrows to Cedar Cove and the lower road had been washed away.[29]

There were a lot of people listening to the police broadcasts that night. A fireman picked up news of high water on his police scanner and told Roger Rogers, his dispatcher at the Loveland Fire Department. Rogers alerted the department and requested the volunteers at Drake and Cedar Cove to offer assistance. There was no answer to his call at Drake, but Chief Leroy Graham responded from the Cove.[30] Graham also tried unsuccessfully to phone Drake. It was a mystery; why was there no answer at Drake? The river was normal at Cedar Cove. The Riverbend Motel, just to the west, placed a call for help and Graham went to investigate.

The chief, an older man, had just finished a training course in Texas on disasters and knew something about the problems. He stopped by the fire station, got the truck and started up the canyon. Chronologically, he was ahead of Terry Urista. There was no traffic on the road. He passed the power plant; things looked all right. He crossed a bridge, no trouble. Then, sixty

The debris carried by the current lodged in the concrete guard rail.

Reports of this house lodged atop a bridge near Cedar Cove gave the Sheriff's deputies their first indication of the magnitude of the flood.

yards farther, dark, rolling, surging water struck him straight on. It spun the truck around and turned it on its side. Graham jumped out and as the powerful river nudged and shoved the truck, the aged man with his hip boots filling with water ran for the mountain slope. He made it. The only part of the truck found later was the rear axle.

As he sat on the mountainside, wet from his race with the river and probably only a few hundred yards from Purdy, Graham listened to the noise of the water and worried about his family. Although he did not know it at the time, the Big Thompson claimed the lives of his wife, daughter-in-law, and two grandchildren. His son, Bob, stranded on the roof of a house, watched helplessly as the river caught the others in the family Datsun and hurled it downstream.[31]

The police and fire units were less than perfect in warning of the impending danger. It took a precious thirty minutes to recognize what was occurring. Probably only half of the people in the canyon got a warning of some kind, and then, often, just moments before the flood.[32] The police—this is an estimate and cannot be exact—managed to warn twenty percent of the people involved in the disaster.[33] The major police alert effort was focused in Loveland and at the mouth of the canyon. This action was considerable and meritorious and they saved hundreds of lives. With more experience the police would have reacted quicker, but even then there would still have been problems.

An assumption by most Americans is that improved technology always means something better. With the telephone, ironically, computer dialing, which would be considered an advance in most aspects, precluded the utilization of the telephone system for mass warning. Twenty-five years ago an operator could have rung a series of phones all at once. Each number must now be dialed separately. There would have to be a list of every number in the canyon—there was none at the time—and someone would have to punch out the seven digits, wait for it to ring, clearly deliver the message, and move on to the next number. It would be an impossible task in the short time be-

Cedar Cove, which embraced comfortable homes with tall, green trees, was obliterated.

Cedar Cove.

Cedar Cove.

fore the river took the lines down. Measured in terms of warning efficiency and lives lost, technology, in this instance, stepped backwards. The computer system might be adjusted, however, for emergency warning. According to Mountain Bell, this is a possibility.[34]

There is another obstacle—a problem of credibility. Often people did not believe the alarm, especially when there was no other supporting evidence (such as rain, a rise in the river, noise, the odor of propane, or floating debris). Some had to be told more than once. The guests at the Canyon Inn, for example, moved only after the second police warning, when they were told that Olympus Dam had broken. This was untrue, but they left.[35] Most people trapped in the canyon, however, received no official warning. Alerted by what was happening around them, or perhaps by a neighbor, they acted to save themselves. Some people were just plain lucky.

One of the last persons through the west end of the canyon was Rube Seilback who, with his family, was headed for the Gaslight Inn at Estes Park. It was the last chance of the season for an evening with a favorite entertainer. Seilback noted that it was darker than usual and, just east of Drake, he ran into light rain. Two miles farther it became heavy and intense. Fifteen, or so, cars had stopped along the road waiting for it to halt. His wife and daughter suggested that he stop too, but he kept going. The mountainside seemed all water and it gushed across the road in a half-dozen places. Seilback swerved and dodged, and plowed onward. He made it.[36] He was among the lucky.

Judge James Kimmel came from the opposite direction and must have passed Seilback. Kimmel, an attorney, judge, and president of the Arapahoe County Bar Association, was driving a relaxing circle excursion which took him through Estes Park and down the Big Thompson. There had been rain in the Park, but it had subsided when he started down. About three miles before Drake, however, he had to pull over with the others. There was a foot of water on the road and rocks were bounding and bouncing off the canyon walls and across the highway. Kimmel picked a spot where there were trees and the slope not too steep. Waiting, he noticed an odd reflection on the road. Startled, he realized it was a car floating by with its headlights on. Other cars and a mobile home slipped past. He could not tell if they were occupied.

This meant there was danger for him too. At 9:00 p.m., in the darkness and downpour, he began to climb uphill with seventy other people among the trees. They could not see much except when the lightning flashed, but they could hear the river behind them tearing at trees, and the rolling hiss of ruptured, floating propane tanks. Until dawn they huddled on the drenched mountain slope and then slogged down to their fortunately unharmed cars. They made a fire from broken furniture on the road and vainly searched for news on the radio.

At 8:30 a.m. helicopters appeared overhead and an hour later a small one landed to evacuate six people. Another did not come until 12:30, and it brought stunning news. A woman leaped out of the helicopter, ran to them, and said she had been officially sent to warn them that Olympus Dam had collapsed. Tons of water were rushing down the canyon upon them. They should run for their lives. Hysteria broke out as the helicopter whirled away to warn others. Kimmel raced back up the slope. In the scramble he met a scared bobcat peering out of a den; the

Cedar Cove.

Cedar Cove.

judge took refuge in an abandoned bear's cave.

Curiously, Kimmel noticed only fifteen of the seventy moved to the proven safety of the mountain the second time. The rest remained below and waited. Why? They knew the terror and power of the river; they had just experienced it. Why wait for death? Kimmel surmised: "It's hard to believe something you don't see. The main problem in a disaster is the inability to react. It takes a lot of confidence to survive."[37] It is true that a warning takes on added significance with surrounding reinforcement. This is the proven case with initial warnings.[38] But these survivors had been through it once, they knew what to do to live, and yet they did not react. Elsewhere in the canyon, others given this same sort of double shock did not move the second time to save themselves. It was a curious phenomenon.

Kimmel, however, with his few companions stayed in the cave, made a fire, sat on pine needles, and caught rainwater in a baseball cap. "I'm really thankful that I was a Boy Scout in Omaha," he commented.[39] The news about Olympus Dam was false. The dam was never in danger of breaking, but the rumor persisted throughout the disaster. It came up over and over through crank calls on citizen band radios and garbled official messages.[40] It was not an illogical assumption, of course. It was believable. People had a hard time accepting the thought that so much water and destruction could come from a thunderstorm. After all, thundershowers are an everyday occurrence. Moreover, the newspapers had told only a few months earlier about the collapse of the Teton Dam. A dam break made sense.

The Olympus Dam rumor and the reaction to it reveal other significant elements. Some of the victims displayed noticeable fright when they heard about the rumor. Panic is expected in the face of danger. Disaster studies show, however, that most people are calm, rational, and resourceful. They may become excited when under immediate threat, but they do not go into a cata-

Above Cedar Cove Captain Terry Urista discovered the road washed away along the outside curve of the river.

The Loveland Power Plant above Cedar Cove disappeared in rock, rubble and water.

The generators of the plant were left in the sand.

tonic state.[41] There is little example of immobilizing shock in the Big Thompson story, other than the curious second shock syndrome that affected some people.

Kimmel eventually came down from his refuge and left by helicopter. Ervin Deal had a similar experience along the highway. With his wife and another couple they drove the canyon road on their way to a square dance near Estes Park. Two miles above Waltonia, in heavy rain, a washout stopped them. A side gully, spewing mud, water, and rocks made it impossible to continue, but a man and woman with a motorcycle tried to walk through. The woman and the cycle slipped over the edge, but she was pulled back. The motorcycle remained lodged against some rocks. About nine o'clock the river began to lap over the road. When the Deals saw a van and a car in front of them wash into the river, they stepped into the rain and started up the cactus-strewn slope with their companions in their square-dance costumes. After climbing for three to four minutes, they turned to see their car, illuminated in the headlights of other automobiles, slide off the road and into the river. They never saw the car again.

Cold in the soaking rain, they remained on the slope for two hours in pitch blackness. A man with a flashlight finally led them down and they took refuge in the back of a van which contained some sleeping bags. They spent the night there. At ten the next morning, a helicopter arrived to take some of the group to safety. At 12:30 p.m., a helicopter arrived with the same news Kimmel received—the dam had broken. Amid screams and excitement, the people scrambled back up the hillside. No more water appeared, of course, and the Deals were eventually taken out in mid-afternoon.[42]

In the same general area a band of thirty-one stranded motorists formed a human chain which hand-in-hand reached the shelter of a cabin. The people, including children, quietly huddled there and allowed the water to surround them.[43] The cabin, fortunately, did not float away. Why did they wait? Why not go back into the storm and climb higher where it was safe? Perhaps, because of fright and the presence of temporary shelter, the compulsion for a second effort was not present.

A short distance away from his front porch at Glen Comfort, Ed Renner watched the events across the river at the Stockdale Motel with horror. His parents owned the motel and his mother was there with his brother's girl friend, Tammy. Shortly before the river washed the bridge out his brother, Kenny, had crossed the bridge to his side. A tree had fallen in Ed's yard and knocked over his mail box; Kenny thought he might be able to help. Now he could not get back. Renner's new car and Kenny's pickup truck, parked at the Stockdale, began to dance around as the water moved up to the motel. With the other cars, belonging to his parents and motel guests, they formed a brief traffic jam in the river before moving downstream.

The lights went out which upset the children in the house with his wife, Darleen. "Mama," said Renner's eight-year-old daughter, "I know we're all going to drown, and I know we're all going to be in heaven together, but will we have to bleed?" He tried to calm the children and watched helplessly as the water rose against the Stockdale Motel. When it reached half-way up the windows, the candle in his mother's room flickered out.

Remmer waited as fierce horizontal lightning cracked through the canyon and the repulsive odor of propane filled the air. The storm subsided and he ventured out. Finding a concrete bridge intact, he crossed the river and worked his way around to the back of the motel. There he discovered sixteen people crowded under wet blankets including his mother, Tammy, and the pet Siamese cat, Misty. As the water invaded the lower floor of the motel, a guest had cut through the ceiling using a knife and hammer. He shoved the others through the hole and they all escaped outside from the upper story.

Renner led them to the shelter of his home and explored some more. He found three people who survived by clambering up into the rafters of their cabin. Two couples had been visiting when the storm hit. One of the men, Edward Morton, went outside to move a pickup truck to safety, but he was caught and did not return. His body was later found at Renner's ruined bridge, still in the vehicle. The others retreated to the ceiling beams and remained there until the water went down at 3:00 a.m. Luckily, the house withstood the moving water. At 9:30 a.m. the National Park Rangers arrived and led the survivors out on foot.[44]

Gary Haxton and his family decided to abandon their cabin. He had been to Estes Park for his daughter's wedding in the afternoon, and had planned to spend the evening with his wife and other children at the family cabin at Waltonia. Located on a gulch behind the S.S. Rapids and Waltonia Motels about thirty feet from the river, the vacation home had a screened sleeping porch overlooking the Big Thompson. When Haxton went to bed, between 8:00 and 8:30 p.m., it was raining hard with thunder and lightning. Near nine o'clock the lights went out and he looked at the river. He saw cars moving on U.S. 34; it was all right. When he sat up a few minutes later to check again, however, to his astonishment the water was almost as high as the porch.

Some neighbors wandering outside to observe the river warned them to leave at once. As they fled out the door in pajamas, a large pine tree by the cabin twisted, snapped, and fell with a splintering crack. They were told that Olympus Dam had broken. They hastened uphill. They were taken in by David and Maybelle Leathers who lived higher in the gulch. The Leathers had lit candles so that others might see the light and take refuge.[45]

When Haxton and others went out at dawn, they discovered that the cabin remained, but nearby motels were gone. To their surprise they found a dazed man standing on a small island in the stream. They chopped down a tree to form a crude bridge and helped him to the bank. His hands had been ripped and torn like hamburger. He was cold, in semi-shock, and told a wild story.

He was a tourist visiting from Iowa with his brother-in-law. He had stopped at the S.S. Rapids to make a phone call. Separated from his wife and children who were traveling in a different vehicle, he was trying to locate them. By the time he finished his call, the water was chest-deep. With ten other stranded motorists, he swung onto the roof of the motel. Under the pressure of the water, the motel fell apart. As it collapsed, he grabbed a tree standing nearby. His brother-in-law seized his leg, could not maintain

At Drake the flood left a bed of rocks.

This area of Drake was once covered with small houses. The highway maintenance building survived.

his grasp, and fell away into the maelstrom. Alone and surrounded by terror, the man clung to the sturdy tree through the night, descending when the water receded at daybreak to the small island.[46]

The night before, when the water began to overlap into the parking area, the manager of the Waltonia Motel told people to leave. Most of them did and spent an uncomfortable night under a ledge on the mountainside.[47] From a cabin close by two young men, Ken Bohling and Kurt Koerner from Arvada, Colorado, helped people escape the river. Ken assisted some travelers leave their floating car, got hit by a log, and spent the night shivering on the mountain wondering about Kurt. His friend had disappeared.

During the evacuation the next day, which Bohling helped organize, he was told that a nine-foot wall of water was coming down the canyon and that there was not enough room on the helicopter for all the survivors. He elected to remain behind with an old woman, and then, strangely, he went with her to her cabin to await their fate. It was the second shock syndrome. When asked why he had not gone to high ground, he merely replied, "I couldn't really think straight anymore. I was pretty shook up."[48] Ken Bohling, of course, lived; there was no nine-foot wall of water. Unfortunately, his friend died. Somehow, unseen, in the blackness, the swift water took him.

Along with Kurt Koerner, one of the honored heroes of the flood drowned at Waltonia. This was Officer Michel O. Conley, a summertime policeman at Estes Park. Off-duty, he was traveling with his wife through the canyon when the water stopped them. He told her to go to high ground while he remained behind to assist those in trouble. As he was trying to help, he was entrapped by the rising river. Running, he seized hold of a light pole, but it snapped and he was gone in the muddy water.[49]

At the same time, in Glen Haven (on the North Fork), there were similar problems. Eerily, a

The Glen Haven Lodge where hotel guests helped the owners battle rain water leaking through the walls.

medium in Estes Park predicted dark clouds and blackness around Glen Haven while reading her client's cards.[50] It was a strange evening, indeed. Gray-haired Dorothy Ferguson, a full-time resident, noticed it. It had been foggy, then there was hail, then rain. The lights failed at 9:15 p.m., but the lightning revealed the rapidly rising water. She observed neighbor Ernie Conrad's bus floating by and the house trailer across the road moving over her lawn. She never cared much for the mobile home anyway, but it knocked loose her propane tank and sent it hissing into the stream. The water swirled around her house—she lived on the inside of a curve—so she waited and watched by candlelight. Her backdoor led to the mountainside. She determined that when the water reached her front step, she would go out the backdoor and climb. She was safe, for the moment.

A zealous local fireman arrived, however, and physically forced her to evacuate. He put her in his jeep and with others aboard promptly drove off into a large hole filled with muddy water. Dumped into the water and unable to swim, Mrs. Ferguson grabbed a bush to save herself. Dripping, she trudged back home to dry off. Disgusted with the whole business, she went to bed and slept until the next day.[51] It was a bad night.

Close by, at the Glen Haven Inn, owner Bill Wells also noticed the rain, which started between 6:00 and 6:30 p.m. It began as a normal afternoon shower, but it did not quit. He had forty-five guests at the Inn. Water began to seep into the kitchen. Wells and the future owner, Ted Hanes, who was there learning the business, mopped and continued cooking. The wind howled around the building and rain began coming through the walls. They could not stop it, nor mop fast enough. The electricity went off. Looking outside, Wells was surprised to see the North Fork out of its banks and flowing down the street past his front door. Propane tanks and automobiles skimmed down the road. The town hall, across the street, floated off its foundation.

The guests inside the seventy-year-old inn could barely hear above the wind and thunder. They were uneasy, but someone went to the piano and, with wry humor, began to play "Stormy Weather," "Old Man River," and "Moon River" in the candlelight. They all pitched in to catch water in roasting pans and sop up seepage with rags. They shared rooms in good spirit and when the rain subsided, they surveyed the damage. "Jesus Christ!" gasped a visiting school teacher when she saw the parking lot at daybreak. The cars were covered to the windows in silt. The inn had no water, no toilets, no electricity. The food was spoiling and the kitchen was a mess. When they got the opportunity, the people left. The deserted hotel then became the headquarters for rescue and recovery.[52]

Two miles downstream, Walter A. Skowron lived in a cabin on a hill. The North Fork there was normally six feet wide, but with the heavy rain it spread sixty yards. He and a visiting friend, Jerry Keenan, noticed the early nightfall and the enlarging stream. They smelled propane and listened to the noise. It sounded like a Mack truck grinding its gears and roaring past. With increasing apprehension they turned on Keenan's C.B. radio. They were stunned at what they heard. The channels were jammed with distress calls. An unknown person pleaded over the airwaves, "I'm trapped in my van with my wife and

Drake.

two children. For God's sake, help us! Ohhh."

Reception is short distance in the mountains, so they knew the calls came from near at hand. Shocked, they agreed not to tell their families in the cabin the news, so that the children would be able to sleep. But they stood watch all night. The next day they hiked to Glen Haven through debris and fallen trees. They killed a rattlesnake on the way. After several unsuccessful attempts, they were able to find space on a crowded helicopter for their families. Skowron's mother had a heart attack while going up the hill to the evacuation point, but they got her out safely.[53]

Farther down, where the North Fork joins the Big Thompson at Drake, Officer Littlejohn had his share of difficulty on the night of the flood. He turned some drivers back down the canyon and warned others. Jim Shoots, a quiet bus driver in charge of thirty-five teenage students and nine faculty advisers from New Albany, Indiana, needed no official warning. He turned back on his own. They were in the canyon on their way to a rodeo after completing the annual school mountain climb of a 14,000-foot peak. He noted the condition of the river and simply told the chief faculty member, "I don't like the looks of this." He stopped, turned the bus around, and at 8:45 p.m. got out before the flood struck. He saved those in his care by intuition and good judgment.[54]

Once the rise in the Big Thompson became noticeable, Littlejohn received no argument from the people he turned back. They followed his orders and left. He moved his patrol car to higher ground. In the darkness he met Gordon Kindred. The old man had gone outside to move his jeep to the roadway. Before he could return to the cabin, where he lived with his wife and elderly aunt, he was cut off by the water. The encircling river took the cabin. Littlejohn had to forcibly restrain Kindred to keep him safe. "You just had to see it and you knew that if you got caught in that water, you were dead," the officer observed. "It was like boiling, muddy water down a dry gulch with boulders that sounded as big as my patrol car; houses and cars floating like corks."[55]

Drake.

The grinding stones of the river demolished automobiles.

Drake. *Photograph by Mary Alice McComb.*

Exploring when he could and wearing out his boots in the water, mud, and stone, the patrolman stayed the night at the confluence. He located the foundation of Kindred's home, and later found a place for him to stay. He met a girl at Drake who had slept through the whole affair, pulled the master switch for an old couple whose house had moved three feet off its foundation, and, unfortunately, discovered bodies in the debris.

Littlejohn and his car became a focal point for the helicopter evacuation. Although he had followed orders and acted with courage, the patrolman still felt a certain remorse.

> I'm grieved when I think I didn't save more people, but how could I imagine what was coming down. I had trouble getting people to believe the feeble excuses I had, much less something of that magnitude. Now, I regret turning people back down the canyon, because we all know that anyone caught in the canyon had no chance. There was no way to foresee where the danger zones would be. Nobody really knew until it got there.[56]

Robert Hicks was really caught by surprise. He had gone to his car at Drake to hear a football game and thereby missed all of the warnings. As he listened, he saw something move outside. He opened the door and, to his astonishment, water poured in. He looked at his house; he couldn't make it. Worse, as he contemplated his position, the car lurched and began to move. There was little time, only seconds. He saw a pine tree standing in the course of the floating car, and poised on the open doorframe. As the car slid underneath, Hicks leaped and swung out on a branch. He climbed up and then hugged the tree for hours in the wind, rain, and thunder.

A cabin upstream topped over, moved, and with roof pointed into the current, stuck in a row of trees. The overturned cabin diverted the flow from Hick's refuge and his house. His wife and daughter could see him in the lightning flashes clinging like a frightened bear cub to his pine tree. Propane tanks sailing by were pulled under by the boiling water. They would wetly pop to the surface and, with gas hissing, sound like spouting whales. A sixty-foot house trailer rushed by. Hicks briefly saw a screaming man and woman waving a flashlight in the doorway. The raging Big Thompson dragged the trailer underwater and destroyed it.

Hicks remained in his tree until the water receded and he could wade to safety.[57] Virginia McGee had more choice than her neighbor and did not wait to find out what might happen. Three months earlier she and her husband, Ross, had leased the profitable Drake Stage Stop. It was the main store, gasoline station, and refreshment stop for tourists. She had heard about the rocks on U.S. 34 at Waltonia and called the Sheriff's Office. Blocked motorists began to collect at the store and, in a short while, Littlejohn rolled up to warn people to leave. His message, garbled by the public address system in his car, was unclear, but his meaning became apparent when Virginia saw the North Fork reach the bottom step of her back porch. The Big Thompson had not crested and people began to leave.

The McGee's scrambled for their car and headed for high ground at Cedar Cove where they had lived for two years. Ross assured his wife that they would be all right and tried to calm her fears. They made it, but it was close. Later, they found black silt and sand in the hubcaps of the car. They stopped at the Park Terrace Motel, high above the river. There they were told that the dam had broken at Estes Park. With that dreadful news, Virginia gathered her children and hustled barefoot up the dark mountainside behind the motel to wait for several hours.[58]

While the McGees were racing the flash flood along the highway, fourteen-year-old Cindy Miller talked over the telephone to her friend, Stannis Bashford. Both worked at the Covered Wagon Restaurant outside Cedar Cove. Cindy had left work early that night. From home she called to warn her co-workers about the condition of the river. The manager thought that they would be all right. Stannis, however, said the water was all around them and that propane tanks were going by the windows. Abruptly the phone went dead. Cindy ran and shouted to her

Drake. *Photograph by Mary Alice McComb.*

The force of the river and the depth of the water could be measured on the trees.

Debris piled along the road to Glen Haven from Drake.

father telling him about the phone, the restau-, rant, and the seven people there. An eyewitness later reported hearing the screams of waitresses as a wave of water smashed a nearby house into the Covered Wagon shattering both structures into the torrent.

Adam Miller, Cindy's father, hastened his family into their station wagon and sent them across his fifteen-high-foot bridge. The water was then three feet from the top. He paused to warn his neighbors and then followed the station wagon in his pickup truck. In the two vehicles, the Millers raced down the canyon and through the Narrows. As he sped toward the State Patrol barricade and safety, Miller noticed an ambulance at the mouth careening past him into the canyon. Miller was likely the last person to exit the mouth of the canyon.[59]

While the water continued to rise at Cedar Cove, across the river from the highway Marj Stengel and her daughter listened as someone on a bullhorn yelled, "Get to high ground!" Urista had been heard after all. When the lights blinked out her husband, James Stengel, went to check Cedar Creek. He found the nearby small tributary two feet over his bridge and still rising. He retreated and kept watch. Later, he noticed a dim light and calls for help from the other side of the creek where D. K. and Ginny Pulliam lived.

When the flood reached their glass terrace doors, the aged Pulliams retreated to their loft and clung together as the river pummeled the house. After a half-hour battering, the home gave way. As it broke apart, a roof beam catapulted the couple into the water. Clinging to one another and clutching a fading flashlight, they floated on floorboards. They were not swept into the mainstream and they managed to grasp a standing cottonwood tree. They climbed into the branches. They were forty feet away when Stengel heard their faint cries above the racket of the river.

Stengel found some television cable, cut it with rocks, and tied it securely to his gate. He called to the Pulliams to "hang on," and waded out into the waist-deep water. Reaching their cottonwood, he tied off the loose end of the cable, and helped his neighbors to safety and a warm fire.[60]

Meanwhile, across the river on the highway, other people gathered. Richard Huffsmith was hosting a church picnic when the power went off. He received a warning telephone call and, with a flashlight, he could see the river coming up. The six couples with him left immediately for Loveland and passed a speeding ambulance coming from the opposite direction. Huffsmith stayed behind. From the top of a hill he watched the front of the flood surge through with a rapid rise. He saw floating propane tanks, heard the grinding of rocks, and watched helplessly as a car, lights on, skimmed by with screaming people on top. Sadly, the Big Thompson claimed Huffsmith's own nineteen-year-old son who had gone to Estes Park for the evening. Somewhere upstream, the river caught him and washed his body home to the bridge at Cedar Cove.[61]

George and Rachel Morgan who owned the Park Terrace Motel located above Huffsmith shared the same kind of worry. Their son, Greg, had also driven to Estes Park for the evening. He had left earlier and made the trip safely despite washouts and rocks bounding across the road. Greg spent the night in the park and could not get word to his parents for a day and a half. The Morgans prayed for his safekeeping and wept when they heard that he was all right.

The Morgans generously offered their facility to their neighbors without charge and housed fifty to sixty people. The rustic motel was high on the mountain slope and fully intact with a working water well. The regular guests had no idea what had occurred and one lady apologized for shorting out the electricity with her hair dryer. At first the travelers thought they could drive out the next day, but they soon learned there was no highway. They had to leave their cars with Morgan until later and depart with the rest of the people in the helicopters.[62]

The two Sheriff's deputies, Terry Urista and James Garcia, who were marooned at Cedar Cove, roamed up and down the short remaining stretch of road during the floodtime. They helped where they could and transmitted radio descriptions to their headquarters. Urista also observed only a rapid rise in the river, no wall of water as was later reported in newspapers. Garcia rescued an elderly woman from the water and risked his life checking a car. He tied a rope to a

Propane tanks floated for miles downstream. Contrary to rumor they did not explode.

The road to Glen Haven from Drake.

tree on the bank and ventured out into the stream. The automobile was empty and the policeman returned to land. The tree he had used as an anchor then toppled into the deluge. The officers saw a mobile home and a car rush past in the river. They spotted and then lost victims alive, but entangled in transient heaps of refuse piled by the cascading water.

At dawn Garcia discovered the half-body of a woman ripped apart by the river, and rescued an old man, alive but buried waist-deep in silt. "Well, it's about time," the man admonished the startled policeman. Caught in his house with a faithful black dog nearby, the old man had stretched out as far as he could to keep his head above water. It reached his chin.[63]

No one experienced the flood in a more terrifying manner than the two ambulance drivers who sped into the canyon as Miller and the church members from Huffsmith's escaped in the opposite direction. The Narrows serve as the neck of a funnel through which the Big Thompson passes on its way to a juncture with the South Platte on the prairie. If ever there was a "wall of water," it would be here.

John McMaster, a heavy-set man of medium height, ran a private ambulance service in Loveland. On the night of the flood, he received a call from the Sheriff's Office to proceed east of the power plant in the canyon. Although he and his partner, George Woodson, did not know anything specific, this was not unusual. They started up U.S. 34 in their van-type vehicle and met the State Patrol roadblock at the mouth of the Narrows. Misinterpreting the policeman's signal, they thought the policemen waved them on. The police were really trying to halt them. McMaster and Woodson hurried on without stopping. In an ambulance service, quickness and perseverance are desirable qualities.

Several cars traveling at high speed with flashers blinking hurtled past in the opposite direction. It was dry; there was no rain. Woodson, playing with the spotlight, noticed the unruly condition of the river. They became apprehensive and when the river began to slosh over the

The water reached the second floor of this riverside house.

highway, McMaster decided to turn around. As he began a U-turn on the road, suddenly, time stood still. Bearing down upon them, only yards away with the noise of a freight train, was a dark wall of water. "In those few seconds before the wave hit us," McMaster reported, "time actually froze. The water seemed to poise over us for an eternity."

It lifted the van fifteen feet up and slammed it momentarily into a V-shaped ledge on the sheer rock wall of the canyon.[64] McMaster remembered:

> It's very hard to describe the terror one feels at that point, because I recall myself being in hysterics. I grabbed the microphone and hollered that we were trapped in the Narrows. . . . The thing that was going through my mind was, 'Make it quick. I don't want to die, but just make it quick.' And, at that point, I knew I was dead, that this was it, and telling them on the radio that we were trapped in the Narrows. I was getting ready to tell my wife I loved her, because I figured this was it, and I happened to look over. And I saw George going out the window.[65]

In the few microseconds allotted them, both men scrambled out the window onto the precarious safety of the ledge. As if thwarted, the angry torrent hurled the ambulance, headlights pointed upward, across the chasm and cracked it into the solid stone wall. Bits and pieces were later found scattered over the Loveland floodplain. On the ledge, the two men breathed air heavy with dust. The dust was stirred up by the air moving in front of the wave of water which struck them.

They began to climb in the darkness. McMaster, who was behind Woodson, slipped, fell, and reached wildly for something to hold onto. Scratching and clawing the stone, he plunged downward toward death. With his arms embracing a large rock, however, he jolted to a stop, his body waist-deep in the current. He was unable to move, he had severely cut one arm, but there he clung.

Woodson, alone above, called and discovered what had happened. He could not see and dared move neither up nor down in the darkness. In the

A side gully undermined this cabin.

confusion, he dropped the hand radio transmitter he carried. He yelled and McMaster blindly stuck up his hand. The radio smacked into it; he caught it. A miracle! The small, wiry Woodson then tried ripping his clothes to make a rope to reach his injured partner. He succeeded, but the cloth was not strong enough to pull McMaster up on the ledge. They waited and listened to the small radio. They could hear incoming calls, but they were unable to broadcast out because of their nearness to the wall. It began to rain.

They talked to each other and joked over the noise of the river. Maybe they should pray, suggested Woodson. "What if God is a Unitarian?" cracked McMaster. They flicked a cigarette lighter in the thin hope of signaling the helicopter they heard overhead in the blackness. A few hours after midnight, the water level dropped and a segment of roadway appeared below them. Woodson scrambled down and they both staggered out onto the wet asphalt. The radio worked; they made contact.

The order came through, "Get back on the wall!" Another surge of water was coming. They retreated once more to their wedge where, braced against a fissure of rock, they clung together for safety and warmth. Woodson had lost some of his clothing in his attempt to make a rope. They were both wet and Woodson began to suffer hypothermia as the water began to rise again.

They survived, however, and with daylight again descended to the patch of road left by the receding water. Woodson fashioned a splint for McMaster's arm and scrap styrofoam shoes for himself. As a junior high school coach, he curiously noticed almost every kind of athletic ball collected in the rubble around him. McMaster's arm, which was cut deeply but not to the point of permanent disability, began to ache as a helicopter lifted them from the Narrows. Their great ordeal was over.[66]

Eastward, beyond the mouth of the canyon, firemen and policemen warned people of the coming flash flood. Even then, death and destruction still occurred. The fifty visitors at the Sylvan Dale Guest Ranch fled. Most of them made it to high ground as the Big Thompson inundated the property, leaving $400,000 in damages and two feet of mud. Maurice Jessup, the owner, and Ron Alcazar, his son-in-law, watched from a hill as the river destroyed their picnic grounds, fish ponds, kitchen, and dining hall. "I felt very desperate and helpless watching all the very beautiful things go," said Alcazar. "It's a very weird feeling that you can't do anything."

Among those who fled Sylvan Dale were thirty-seven members of the Campus Crusade for Christ. Seven of the Christians failed to make it. They lost two cars with passengers; the fate of only one is known. Five women in a Ford Granada trying to escape rounded a curve and splashed into water coming over the road. The current swept the automobile into the stream. "The car rolled and tumbled in the river," said Marilyn Henderson. "It was horrible. Water was coming in. When it got neck deep, I rolled down the window and passed each of the other four girls out. I was the last."[67]

Caught, the torrent swept the women downstream. Henderson swam, grasped a tree, and climbed above the water swirling around her. Two ranchers from the area, Joe Carter and Cliff Moore, with Carter's daughter, Susan, were walking in the vicinity. They observed and commented on nature's ironic method of ending their drought. Susan interrupted; she heard screams for help. Her father focused a spotlight into the deluge and found Ms. Henderson clinging, hysterical, to a tree. Using poles to test the depth, Carter and Moore waded through the waist-deep moving water and carried her to shore. She told them her story and they looked for the others. The men found only one more woman, a hundred yards upstream, also in a tree. A rescue team reached her with a cable.[68]

Another rancher in the area lost his life trying to save livestock. Linda Conner awakened her husband, David, when she heard something "really weird" going on outside. He went to look and saw the rotating red lights and commotion on the highway. He also met neighboring rancher Greg Adams who was walking their fenceline with a flashlight. Adams had been warned, and he was moving his equipment and livestock to high ground.

As the two men stood in the dark talking, they heard "this sound, this tremendous, awesome sound" from the river. They walked down to a pasture and, for the moment, the Big Thompson was within its channel. Then a 300-gallon propane tank traveling at high speed came spiraling by them. Conner recalled, "We were marveling at how fast the river was rising, because it was out of its banks three to four feet in just nothing flat." They could hear trees popping and they ran for their lives back across the pasture.

Conner returned home, roused his family, and loaded them in his car. As he was about to leave, he heard Adams and Adams' son, Rusty, calling for help. They had gone back to the low area to save piglets. "God! Help us! Dave!" they yelled. In turmoil, Conner drove his family to safety, told the Sheriff, and returned on foot. "I could still hear them screaming for me, and the hair on my head just stood straight up. All I could do was just talk to them."

Other people arrived with lights. They could see Adams with his arms wrapped around a tree with the water pounding him on the back. Rusty was in the fork of a tree a hundred feet away.

Above Drake toward Estes Park the Big Thompson destroyed cabins and motels.

The last of a motel.

Above Drake toward Estes Park.

Conner recalled, "I was turned and trying to get this helicopter to land, and some girl put her hand on my arm. And she said, 'I don't see him anymore.' And I looked back and he was gone."

Rusty was still there, with water up to his arms. As Conner brought underwater rescue people to help, the water momentarily dropped. They merely waded through ankle-deep water and brought Rusty to the hillside. The water then rose again. Les Peterson, another neighbor, observed sadly that Adams had "lost his life over a damn hog."[69]

Peterson, who lived five miles west of Loveland, also noted, "Anyone who wanted to get out, could have."[70] The police covered the area with knock-on-the-door, face-to-face warnings. People doubted, they hesitated, and at the Riverview Campground they had to be told three times. Joe Carter, the rancher who had rescued the Campus Crusade girl, explained:

> Take some of the people who have lived here for thirty years. They've watched that river and they've seen it come up so high, then it goes down. They probably figured this was just another one of those. In that case, would you want to drive all the way down to Loveland at eleven o'clock at night?[71]

Ken Williams and his wife Irene who lived in a trailer at the Riverview Campground were exceptional. They left hurriedly at the first warning. During the 1972 Rapid City flood, the Williams had been stranded for five days without food and water. Some friends from South Dakota had arrived at the campground and they had just settled at a card table when the police alert came, shortly after ten o'clock. They immediately took flight and left everything behind. Experience can be a great teacher. "I'm going to look for a furnished apartment on high ground," Williams later concluded.[72]

Sheriff's deputy Gerald Becker had the duty to warn Irving Liss who had lived near the river for eighteen years. Liss also had friends visiting. The lights went out at 9:30 p.m. and they were told to leave. "We weren't especially impressed and kept on visiting in the dark," said the resident. But Becker returned and angrily shouted, "You have to get out now! I mean it!" They left and drove to Loveland. Returning afterwards, Liss found all that remained of his snug home—a strip of sand.[73]

In the Big Thompson disaster, especially in the canyon where there was little warning, the people and their neighbors saved themselves. There were some police rescues, but not many. This circumstance is similar to that of other disasters; people act to save themselves. They are in the midst of crisis and, by and large, act with responsibility and concern.[74] A research study by Eve Gruntfest from the Institute of Behavioral Science at the University of Colorado at Boulder reported that residents had no better idea about what to do than tourists. Over half of the 399 people included in the inquiry said that rising water and rain gave them first warning. Older people were less likely to flee. The deadliest situation was to be driving alone. The best action was to climb the mountainside; the worst response was to do nothing.[75]

In spite of statistics and studies about what action to take, there are always exceptions. Louise Dikeman with four children and her sister-in-law had been on a fishing trip. On their return, the flood left them stranded in their pickup truck on an eighteen-foot strip of highway. They waited in the truck while the river tore at the banks in front and in back of them. Their strip remained and they survived. The next day a helicopter flew over, hovered, and announced that Olympus Dam had broken. They could take only one more passenger. Dikeman wrote a hasty note, pinned it to the clothing of her infant son, kissed him, and in an act of maternal sacrifice handed him on board. The helicopter whirled away to safety in Loveland and left the others to their fate.

There was no break in the dam, and after several hours, two Park Rangers trudged down the muddy slope from above to lead them out. Twice, the women had waited instead of climbing out. Maybe, they thought, they could not manage the mountainside as the Rangers had done later. "It was more than just luck that we survived," Dikeman explained, "it was divine intervention."[76] Perhaps, it was.

Automobiles were left stranded.

Chapter III

Rescue: The Quick and the Dead

Captain John Englebert, Sheriff Watson's Paul Bunyan of search and rescue, had finished eating dinner out with his wife. He was on the highway headed toward their home in the mountains north of Fort Collins when he heard about the trouble. The police dispatcher told him about the people trapped at Grandpa's Retreat in the Big Thompson and the need for emergency crews. The captain asked the dispatcher to order the motorized patrol to the mouth of the canyon while he and his wife switched to a county-owned pickup truck parked at his home. The truck was equipped with several radio systems—Sheriff's, fire, and telephone—and he used it for field communications in the emergencies he dealt with. Since Joyce, his wife, was also a Sheriff's officer and could run the various radio nets, she went with him.

As they drove south toward Loveland, Englebert could see black clouds and intense lightning in the backcountry. He tried to estimate the magnitude of the problem and listened to the reports of Urista and Garcia at Cedar Cove. When he heard his fellow captain describe the A-frame house which had washed atop a bridge, he knew that he was into something big.[1]

Urista had also requested an alert of the motorized patrol. In response, Rich Hovey, the captain of the volunteers, rallied his people. With several members of the mountain rescue team, Hovey started down the same road as Englebert with "full lights." The dispatcher instructed Hovey to caution the residents east of the Narrows and he began at once to knock on doors, shout a warning, and move on. He did not pause to explain or note a response.[2] The Loveland police and other Sheriff's deputies joined in the call to evacuate the low areas around Loveland.[3]

Captain Thomas of the State Patrol was also enroute. His dispatcher in Greeley had told him about the mudslide at Waltonia and that Miller had gone to investigate. At 9:20 p.m. the dispatcher called again to tell Thomas that he possibly had two men trapped in the canyon. Worried, the captain dressed, got in his car, and drove west to Loveland from his home in Greeley. In thirty minutes he pulled up at the Dam Store, a tourist stop at the mouth of the canyon. The roadblock was in place, Sheriff Watson was waiting, and Englebert wheeled his pickup truck to a halt behind him. Thus, the principal organizers of the search and rescue operation were all there—Watson, Thomas, and Englebert—they met together on the highway amid the illumination of flashing red signals and the distant rumbling of the Big Thompson storm to discuss strategy.

The rain had not reached them. It was dry; so dry, in fact, that the men could stamp their feet and stir up the dust. At the moment, the river was normal and the blocked motorists were vocal and frustrated. But suddenly the water began to rise. They could hear propane tanks clanging, rolling, and fizzing in the growing turbulence.

Captain John Englebert took charge of field operations for rescue and evacuation. *Photograph courtesy of Loveland* Reporter-Herald.

Together, John Englebert and Robert Watson formed a team to handle the rescue. Sheriff Watson directed overall operations and handled political aspects, while Captain Englebert managed the actual evacuation operation. *Photograph courtesy of Loveland* Reporter-Herald.

Officers directing the rescue and evacuation first gathered at the Dam Store at the opening of the canyon.

They could smell the gas.[4] It began to drizzle and Englebert pulled on a raincoat. The men talked some more and ordered a patrol to the Poudre Canyon to the north. If all the canyons were to flood, they might lose a city. They needed more information, and they had lost contact with Sergeant Purdy.

In the darkness, something heavy struck hard against the supports of the siphon. Close to a million pounds of steel and freighted water broke, crashed, and rolled 600 yards down the river. It crunched into the end of a house.[5] Water from the western slope spewed across the Narrows at 500 cubic feet per second in an impressive nine-foot diameter plume. The Sheriff called the Bureau of Reclamation. They doublechecked and then shut down the siphon. It took them an hour. Earlier, the Bureau had closed off the discharge from Lake Estes into the Big Thompson because of the heavy rainfall.[6]

Along with the siphon, U.S. 34 in the Narrows "exploded" and fell apart. Large chunks dropped into the boiling dark water as the chasm returned to a primordial state. Strangely, a woman's voice came over the Sheriff's radio network saying her boyfriend had been swept away. It was curious. Why was her voice on the Sheriff's radio frequency? It added to the confusion, but there was little time to consider its origins. They could hear a roar echoing down the canyon walls. It was time to leave, to get away from the mouth. According to Captain Thomas, there was no sense in having to rescue the rescuers. They began to retreat.[7]

Rescuers elsewhere, however, did run into trouble. At 9:30 p.m. Saturday night in Estes Park, Lieutenant Leo Baker from the Sheriff's substation asked the National Park Service at Rocky Mountain National Park to help clear rockslides in the upper canyon. Most of the heavy equipment was on the western side of the park, some distance away, but the rest was on its way in twenty minutes. By word of mouth, even on a Saturday night, the men heard about the problem and appeared out of the night, ready to go to work. They were professionals.

It was raining with increasing intensity as the twenty-five Rangers moved into the upper canyon. They hoped to spread the washes to broaden the flow of water and repair the road as they went. They had portable radios but necessarily communicated through their own dispatcher who, in turn, relayed messages to Englebert. Baker's contact with Watson—from Estes Park to Fort Collins to the field headquarters—was also circuitous. The Rangers penetrated to Glen Comfort and, in the rain and lightning, gathered people for evacuation. The road, however, washed out behind them and they were trapped. With the water on the rise, they strung ropes up the side of the mountain, in case they had to climb.[8]

Trapped with the group were Fred and Edna Woodring in a four-wheel drive International Scout. They lived in Estes Park and had heard a distress call from the canyon over their C.B. radio. They drove in to see what they could do to help. At the washout, Woodring attempted to plow through the water cascading off the mountainside. Over a C.B. in Estes Park horror-struck listeners heard Fred say, "It's too much, the water's too high. We've got to turn back." The plunging water pushed the Scout whirling into the river, headlights skyward. Edna blurted over the radio, "My God! It's the end of the world!" The vehicle turned upside down with Edna caught inside, where she died. Fred Woodring managed to climb outside, however, and he clung to the car as it was washed over by debris and the current. David Essex, the leader of the Rangers, tried to throw a rope to him, but Essex was unsuccessful. In moments, the swirling water carried Woodring away to his death.[9]

There were fifty people stranded with the Rangers. They remained there in the driving rain for the rest of the wild night. They could smell the propane gas which bubbled and hissed from the bobbing tanks, and periodically they could see the nightmare world of the Big Thompson in a momentary blue-white blaze of lightning. More Park Service equipment rolled up on the opposite side of the washout, and finally, in the morning they got across.

The Rangers then worked on the road and scouted for survivors as far east as Grandpa's Retreat. People recognized their uniforms and were happy to follow them on foot out of the devastation. One tough old lady, when asked if she wanted a helicopter ride to Loveland, laconically replied, "I know where there's a bed in Estes Park," and trudged out with the rest.[10] It was not easy to walk through the moisture-laden terrain. Alvin Kruchten who hiked back and forth between Estes Park and Glen Haven on Sunday said that the earth was so saturated that thirty-foot pine trees would not hold when grasped for support.[11]

In the canyon, the men working on the highway heard about the rumored break in Olympus Dam. Worse, the news was confirmed by their own Park Service dispatcher. Ron Cotten who was in charge of the road equipment hopped aboard a dirt loader and started up the rough road at forty miles per hour with his yellow lights flashing. Other National Park people took the same action in an effort to save the equipment. Some of them even backed out at the same speed with an eye cocked for an imagined forty-foot wall of water. As Cotten topped the hill at the head of the canyon and headed down the other side to safety, the policemen at the roadblock scattered like frightened quail while the Rangers roared past.[12]

The National Park Service contributed fifty-four people, who worked the equivalent of seventy-five man-days, to aid in the disaster rescue. In groups of forty to fifty, the Rangers led out two to three hundred survivors and helped repair the road in the upper canyon. They pulled back on Tuesday, and shut off traffic temporarily from their western portal to hold back sightseers.[13] They had reacted in the best spirit of the Park Service and the nation could be proud of them.

From the east end of the Narrows, meanwhile, the Sheriff and the others retreated in the rain to the Merri-Ax on the night of the disaster. The Merri-Ax was a small nightclub on U.S. 34 located on a bluff overlooking the river between Loveland and the canyon. A small observation team remained at the Dam Store. Hovey helped with the retreat and then returned to spend the night. He saw no wall of water at the mouth of the canyon, but heard the grinding of boulders. He was there when the Bureau of Reclamation servicemen arrived to check the siphon and shut down the small power station located there.[14]

The danger of the river forced rescuers to move to a second headquarters at the Merri-Ax.

In front of the Merri-Ax on the highway, Englebert ordered the construction of a heliport. The components, in multi-colored boxes, were laid out with a proficiency—one, two, three—that amazed Captain Thomas who observed the procedure.[15] Englebert then called for the Flight for Life medical helicopters from St. Anthony's Hospital in Denver and the small commercial machines from Ag Helicopters in Fort Collins. Gene Greenwalt, president of the local company, received his first alert at 10:15 p.m. from Earl Denton, the county civil defense director. Every twenty to thirty minutes he received additional alerts from the Sheriff's Office, and finally, at 2:30 a.m. through a light rain, he flew to the Big Thompson. With spotlights shining through the mist he tried to survey the Narrows, but he could not see well and it was too dangerous to continue. He landed to await daylight.[16]

In the darkness Englebert heard from his advance post that a car was coming down the river with its lights on. He ran over to the embankment and, in a few moments, it bobbed by, cork-like, with faces and flashlights inside. There was no time to help and in seconds it rushed past. Some bodies were later found in a car. That night his men found seven corpses in front of an elementary school near the river. He also heard about the two Campus Crusade girls rescued from trees. To everyone's relief, a Salvation Army canteen truck wheeled in, and the owner of the Merri-Ax arrived to open his building.[17]

Other people began to assemble at the command post. A Mountain Bell telephone man arrived in a Ford Pinto with a bank of portable phones; highway representatives came and began to order heavy equipment; and, significantly, the first politician—James Lloyd, state representative from Loveland—appeared. Shortly before ten o'clock while driving the streets of his hometown, Lloyd heard over his car radio reports from a Denver station about the flooding. He drove home, called people he knew who were in the path of the water, and helped a neighbor move his thoroughbred horses out of harm's way. At 1:00 a.m., thinking he might be of use as a

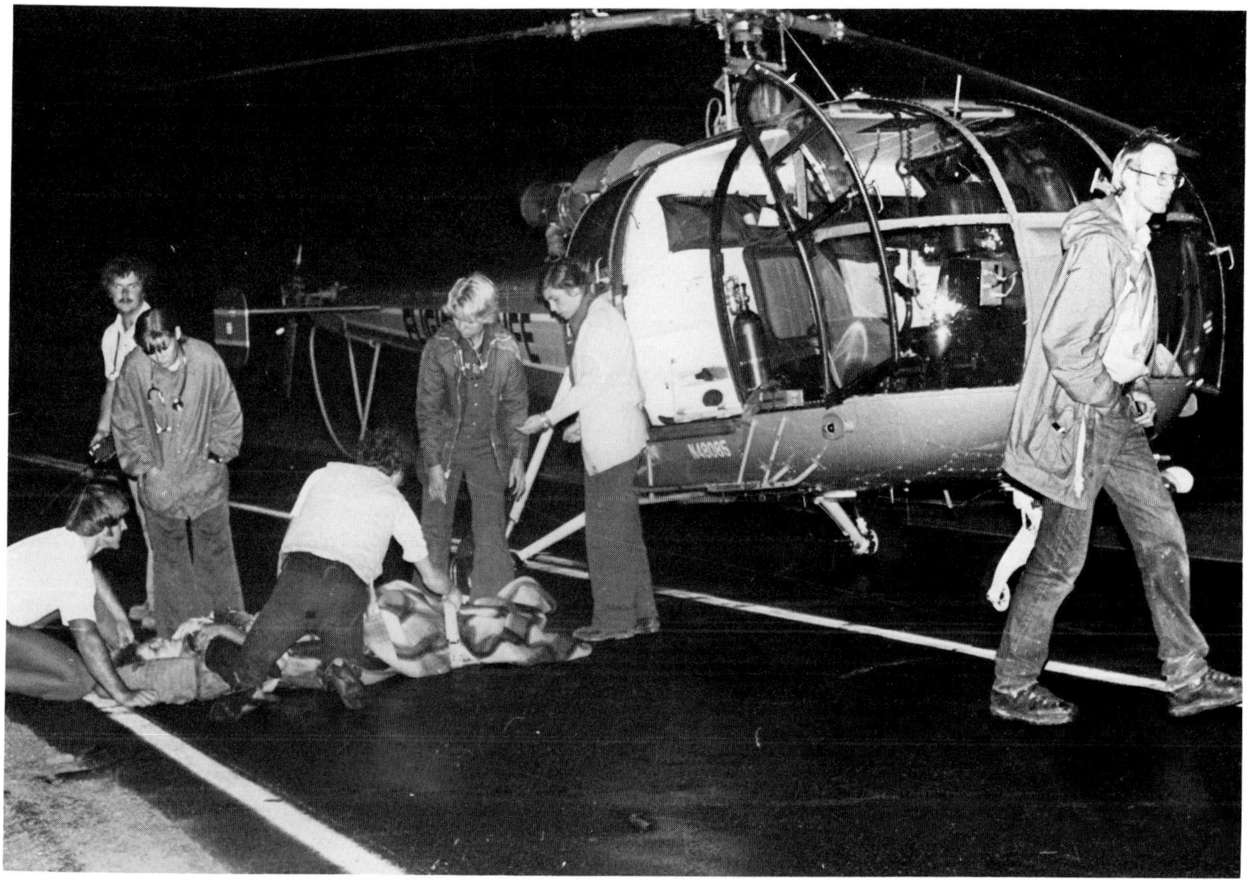

On the night of the flood, helicopter rescue took place at a helipad on U.S. 34 near the Merri-Ax. *Photograph courtesy of Loveland* Reporter-Herald.

liaison between agencies, Lloyd went to the Merri-Ax to offer his aid.

It was crowded and he found it hard to park. The only lights came from car batteries and generators, but he found Englebert and asked if they wanted the National Guard. There was no decision yet; it came later. At 3:15 a.m. Lloyd called the governor's mansion in Denver.[18] Governor Richard Lamm had had an earlier conversation with the State Patrol after he had returned from viewing *The Ballad of Baby Doe.* It was the eve of the Colorado centennial, and he had been to Central City for the play. The next day he was supposed to hike Mount Evans, the state's highest mountain, and make a speech at the capitol. It was a full schedule. Nevertheless, as he went to bed, Lamm told his guard officer to keep him informed.

The governor took Lloyd's early morning telephone call. The Loveland representative was one of the most conscientious of the younger politicians, and the governor trusted his judgment. He told Lloyd that the executive office required an official request from the Sheriff. The young, liberal governor had grown wise in office. He knew well the statutory regulations from an earlier experience with the prisons. Watson called shortly and asked for not only the National Guard, but also additional state troopers. Consequently, at 4:00 a.m., the governor gave General William Weller authorization to move 160 soldiers. Only a few hours before the Fort Collins unit had left for training in Florida. Perhaps it was just as well because the Denver military police unit which was sent instead had been trained in first aid and law enforcement. Lamm then asked his wife, Dottie, to cover the opening centennial celebrations and arranged to visit the canyon in the early morning.[19]

Also at 4:00 a.m. the first of the tactical planning sessions between the principals occurred. The meetings which followed usually included Sheriff Watson, Captain Thomas, General Weller, and someone representing the governor. They convened, when necessary, two to three times each day and made basic decisions about

One month before the disaster the Sheriff's volunteers learned to set up and direct heliports. Boxes with the necessary equipment were ready when the flood struck. *Photograph courtesy of Loveland* Reporter-Herald.

Lack of space and facilities required another move of headquarters to the Northern Colorado Water Conservancy District Building.

the evacuation effort. As dawn approached, a massive helicopter evacuation took shape. The U.S. Army contributed large troop helicopters from Colorado and Wyoming. The National Guard had been mobilized and the Forest Service, again thanks to James Lloyd, provided ground-to-air communication equipment. The Sheriff's Office produced more boxes marked "helipad" and laid them out in triage formation—yellow for injured, green for uninjured, and red for bodies.[20]

Disaster preparation rarely makes provision for a command post.[21] Englebert's truck was the nearest facsimile and clearly inadequate, so Thomas was given the task of finding something better. The answer was right at his elbow. Tugging at his sleeve was a small man who tried to gain his attention while Thomas discussed the problem with an officer. The man was Earl Phipps who directed the Northern Colorado Water Conservancy District. This organization distributed the water of the Colorado-Big Thompson Project to farmers. It had a modern building, with adequate telephones, on the western edge of Loveland. It was perfect and the command post, including the Salvation Army canteen, moved there. At daybreak they were ready for the evacuation. As Englebert summarized:

> The first night, everything was verbal; we were pretty fresh. I would reach out to someone that I knew had the experience. I would say, 'Get me a heliport down the road at such and such a place, and set it up.' And they'd take off; they'd do it. I wouldn't have to worry about it. This was all the result of about six or seven years of training for this particular night.[22]

Governor Lamm arrived at 7:30 a.m. With Watson and other officials, the governor proceeded on a reconnaissance of the Big Thompson Canyon. He noticed the incredible ruination of U.S. 34, the isolated State Patrol car at Drake with its red lights still turning, and desperate survivors waving for help.[23] Alvin Kruchten saw Lamm waving back from the helicopter window as he whirled over Glen Haven. For those on the ground, the appearance of the governor brought

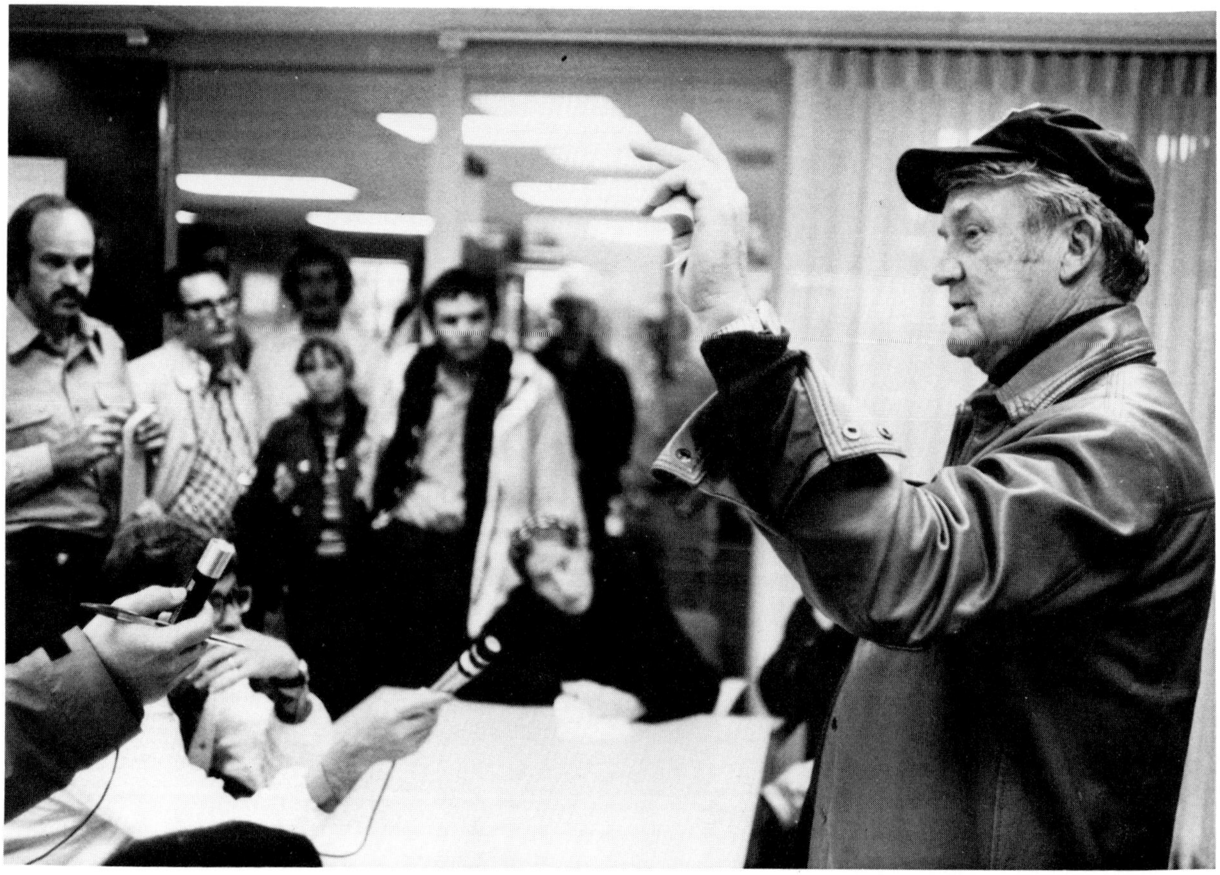

Sheriff Robert Watson possessed clear authority to direct the rescue operation. He handled public relations and met frequently with the press. *Photograph courtesy of Fort Collins* Coloradoan.

no comfort; survivors wanted someone who could give immediate aid, not a politician.[24]

Possibly mistimed, the governor's flight was, nevertheless, a necessary step in the forthcoming recovery phase of the catastrophe. He returned to Denver and asked President Ford to declare Larimer County a disaster area. The President honored the request Monday morning, releasing federal disaster funds for the benefit of survivors and the emergency agencies.[25]

Shortly after nine o'clock on Sunday morning, the National Guard arrived to stay for two days. It cost the taxpayers $40,000, but they were self-contained, disciplined, reliable, and recognizable in their uniforms. There was no question, however, about civilian authority; Sheriff Watson was in charge. The soldiers worked in the canyon and in Loveland and loaned equipment, particularly fresh-water wagons, for use after they returned to Denver. Their helicopters remained in service through Thursday.[26]

Englebert quickly developed a helicopter procedure. The large Chinook aircraft picked up people from assembly areas at Glen Haven and Drake. The smaller craft scouted for survivors and ferried them, a few at a time, to the assembly points. The U.S. Forest Service provided an air traffic control net for the nineteen helicopters whirring in and out of the canyon, and maintenance took place at night under floodlights. It was not an easy operation; flying was hazardous.

The wind force above the Big Thompson chasm was too great, so pilots flew between the walls, up the north side and down the south. Even though gusts sometimes struck at sixty miles per hour, the crews picked people off rooftops, the sides of the canyon, and from rocks in the middle of the river. As one of the Air Force volunteers said, "Nothing is flat up there."[27]

At Thomas' suggestion, the pilots dropped leaflets which gave instructions for self help, laying signals, and remaining on high ground. Incomprehensibly, the Sheriff objected to this action on the basis that it would litter the canyon—as if it should matter. Survivors, some of them at least, read the leaflets and acted accordingly.[28] When the time came to be taken out, people swarmed to the helipads.

The biggest problem at the assembly points was that passengers did not know how to act, and they were more apprehensive about the helicopter ride than about the Big Thompson. The riders, nonetheless, thanked the pilots, said they were glad to be alive, and asked about lost family members. Some residents refused to leave, however, and remained behind. They stayed to protect property and clean up, and because they had nowhere else to go. A six-year-old boy with leukemia also chose to remain. He wanted to die at the family cabin at Cedar Cove, and he met his death there within a few days. An eighty-year-old woman, with cats, dogs, and rabbits to care for, barricaded her door and refused to leave. A group of tourists also refused. They had paid for a cabin for two weeks, and they were going to get their money's worth.[29]

Reluctantly the Sheriff's Office with its helicopters, became the supply source for the canyon. The pilots flew in food, fuel, and even portable generators. At the Park Terrace Motel, a Sheriff's generator served to maintain neighborhood deep freezes. It would have been easier for Englebert to move everyone out and avoid this problem.[30]

On the night of the disaster, the police estimated that there were 2,500 to 3,000 persons in the Big Thompson Canyon. By Sunday evening Englebert felt that seventy-five percent were out. The helicopter operation evacuated 850 people on Sunday; 250 were counted as injured. Of these, however, only 79 needed medical treatment, and only seven had to be hospitalized.[31] Among the worst injured was John McMaster, and among the most unusual was "Baby Moses."

Paul Harvey had broadcast on nationwide radio the story of a lone baby found on a tiny island in the middle of the Big Thompson. The information was false. There was no "Baby Moses," but there was Ronnie Dikeman whose mother had kissed her infant goodbye and sent him whirling to safety aboard the last helicopter in front of the rumored wave of water from Olympus Dam. The note written on a blank check and pinned to his clothing read, "My daddy is in Loveland. His name and address are on the back." In the afternoon the couple came searching for their son, and reunited, left for

During the helicopter evacuation, pilots flew up the right side of the canyon and down the left side. There was only one accident and no one was injured. *Photograph courtesy of Loveland* Reporter-Herald.

home.[32] The true story was interesting enough.

In total, only ten persons required hospital admission. The load at McKee Medical Center was minimal, unlike hospitals of other catastrophes where chaos was commonplace. The hospital activated disaster teams and even sent doctors to the helipads. There was no difficulty, although it would have been more efficient to have better field communications and a heliport at the hospital. Injured people had to be transferred by ambulance.[33] Those in good health were sent by bus to the Red Cross evacuation center at the Loveland High School.

On Saturday evening with increased knowledge of the magnitude of the flood, the Loveland Fire Department asked to open the high school as a shelter. Twenty people from the Riverview Campground stood waiting as the Fire Sirens, a women's auxiliary, started the evacuation center with a gallon of milk, cocoa mix, and coffee. As more survivors arrived, the task became too

Small helicopters lifted people from places of isolation and took them to evacuation points at Drake and Glen Haven. *Photograph courtesy of* Triangle Review.

Large Army helicopters took evacuees to Loveland. *Photograph courtesy of Estes Park* Trail-Gazette.

At Loveland, victims of the flood were given emergency food and shelter. *Photograph courtesy of* Triangle Review.

great. Besides, the purpose of the Sirens was to serve the men of the Fire Department. So they called Barbara O'Reilly, the director of the Red Cross in Fort Collins.[34]

O'Reilly, a good-humored, older lady with a lot of practical sense, had received several other messages including one from the Red Cross division headquarters in Denver. She obtained verification from the Sheriff's Office, alerted her Loveland volunteers, ordered a supply van, and drove to Loveland in the middle of the night. The Red Cross proceeded to serve food, obtain cots and blankets, distribute clothing, screen the survivors, and provide phones for people to use.

In an emergency the local chapter refuses incoming telephone inquiries. If someone elsewhere in the country wants to know about the well-being of a relative they place an inquiry through their own chapter. The request then goes to the divisional headquarters which contacts the impacted chapter in a systematic manner. This allows the telephone system to concentrate on outgoing calls. It is, in effect, a conversion of the lines to a one-way street. Before they converted, however, 400 calls slipped through to Fort Collins. The one-person office was swamped. Afterwards, the Denver divisional headquarters handled 4,000 inquiries. At the gym in Loveland, the Red Cross encouraged the survivors to make a quick call to relatives. They lined up ten deep to use one of the three telephones provided free of charge by Mountain Bell.

At most times there were about 150 people—motorists, displaced residents from Loveland, and survivors from the canyon—at the center during its five-day operation. The turnover was fairly rapid and the Red Cross, along with the Salvation Army, provided clothing, bus tickets, gasoline, and food to help stranded travelers return home. Two thousand offers of transportation and lodging came in to the high school. Gifts of clothing and food were so plenteous that they had to be removed to a dog track for storage and distribution.[35]

It is a curious phenomenon of modern American disasters that donations of food, clothing, various supplies, and even money tend to become overwhelming.[36] It is a by-product of a generous and affluent society, and something which has occurred rarely in the history of the world. Both the Red Cross and Salvation Army keep and store the unused items, in preparation for future disasters. They are not wasted. It is interesting, nevertheless, that for the Big Thompson, the Red Cross spent $134,000 and received back in contributions, $184,000. The Salvation Army gave $20,000 and received $30,000.[37] This fact does not impugn their motivation or good work.

The Red Cross and the Salvation Army have been known to contest for jurisdiction. They both possess recognition as official emergency agencies by the federal government, and both offer the same service. To make matters worse, the Seventh Day Adventists have also entered the competition. At the Big Thompson, however, there was no difficulty. They split the task and worked together. The Red Cross ran the refugee center at the high school and the Salvation Army sent them clothing. In turn, the Salvation Army ran the canteen for rescue workers and the Red Cross made them sandwiches. The Seventh Day Adventists provided lunch for the rescue people, and the Salvation Army prepared breakfast and dinner.[38] There was no prior agreement about the arrangement. It was merely who could do what, and who was there first.

The Salvation Army maintains a large van in Denver complete with coffee maker, portable generator, refrigerator, serving counter, and food. At 10:15 Saturday night the Army's social service center in Denver got a call from the Sheriff's Office and the Loveland police. Bill Birks, a large, blond man with cordial ways, took charge. He picked up additional supplies and reached the Merri-Ax with the van at 11:30 p.m. to begin eight days of serving coffee, donuts, cold drinks, and hot meals. He operated a continual snack bar at the command post and periodically drove around to the various roadblocks to supply the guards on duty. There was no preaching, incidentally. In their own rhetoric, the Salvation Army's contribution was, "No bravery—just

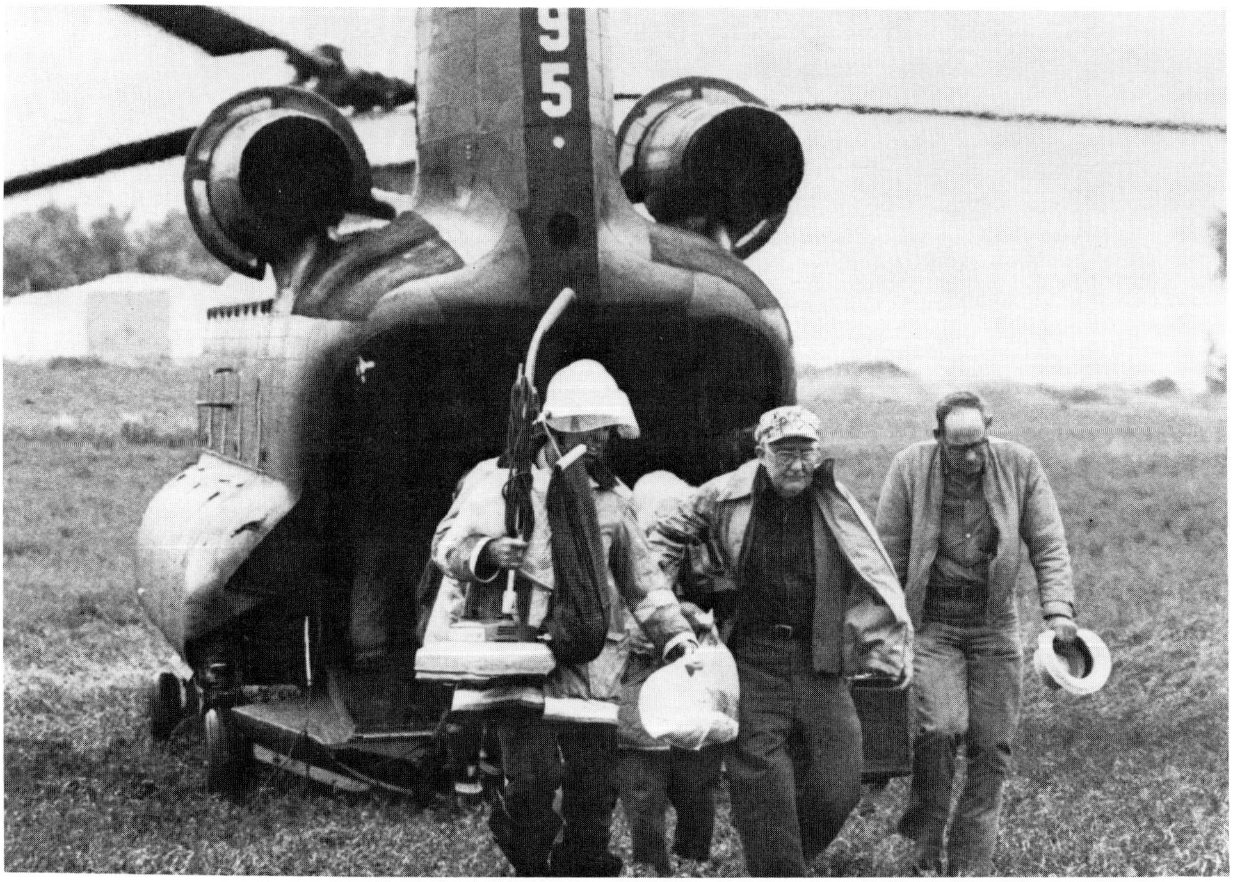

Victims brought out only what they could carry. *Photograph courtesy of Loveland* Reporter-Herald.

Ham radio operators relayed messages to the relatives of survivors and assisted the Sheriff's office by setting up supplementary networks to link the hospital, Red Cross center, and the field headquarters. *Photograph courtesy of Estes Park* Trail-Gazette.

coffee, doughnuts, hot soup, a touch, a smile, a 'God bless you!'"[39]

On Monday, the second day of the evacuation, the weather turned to light rain and mist. It was too dangerous to fly, and so they waited. Everyone—police, newspaper reporters, relatives of the missing, pilots—was restless. Rescue workers checked the river bottoms and the Sheriff organized a fruitless four-wheel expedition to Waltonia. In the late afternoon with a low cloud ceiling, the pilots, flying only a few hundred feet above the stream, returned to the canyon. They brought out seventy-five more survivors, and although they did not fully realize it, the evacuation was over. On Tuesday, when a Chinook helicopter opened its loading door at Glen Haven only a lone dog ran up and got inside. The people either were already out, or had chosen to stay.[40]

The evacuation, which involved 500 officials and volunteers, had been carried out with commendable efficiency. No one had been killed or injured, and no survivor suffered from slowness or ineptitude. This does not mean there were no problems. As Sheriff Watson remarked, "I thought one of the men summed it up pretty well when he said, 'Whenever you get to the point where you handle a disaster too perfectly, you're having too many.'"[41]

One of the worst difficulties was with communications. There was a separate network for the Sheriff's Office, State Patrol, Fort Collins Police, Loveland Police, Loveland Fire Department, Ambulance Service, Forest Service, National Guard, and National Park Service. Various dispatchers had to transfer messages between the systems. The mountain recesses, moreover, were hard to penetrate with a radio. The State Patrol placed a mobile unit near Estes Park, the Boulder sheriff loaned another unit for Estes Park, and Watson had to place a relay station on Palisade Mountain in the lower canyon. This mountain relay station, incidentally, became the scene of one of the more amusing episodes for the police.

On Wednesday, shortly before midnight, the Sheriff's deputies heard a very excited, but fading voice coming over the emergency channel saying something about bears and calling for help. The isolated operator with his radio battery failing was unarmed, so Ag Helicopters dispatched a small coptor to see what was happening. The pilot returned with a shaken, but unharmed deputy who had been encircled in his camp by three hungry bears. The bears loped off into the darkness when the helicopter arrived, but the next attendant went to the station armed with a gun.[42] Reports of other difficulties with wildlife, particularly rattlesnakes, proved unfounded.

Amateur radio operators (hams) relieved the communications problem to an extent. Alerted by the Loveland police on the night of the storm, local operators set up several VHF-FM radio nets at the high school. One linked the helipads, hospital, and command post. Another reached Estes Park to relay messages, and a third was constructed to handle calls from out-of-state. The hams tied into a direct dial telephone system in Denver and utilized an automatic repeater station in Windsor to boost signals. The Federal Communications Commission reserved part of the high frequency spectrum for the Big Thompson disaster. About 150 hams in Colorado became involved. They took care of 1,500 messages in a two-day period. They operated with care and responsibility under the supervision of the Sheriff.[43]

Irresponsible and undisciplined, however, were the citizen's band radio operators. Those at Estes Park were an exception. They kept people informed about the flood and relayed messages from Glen Haven. Even in Estes Park, however, a C.B. call lured the Woodrings to their death.[44] Elsewhere rumors and misinformation filled the channels. Olympus Dam, for example, broke three times on Saturday night according to C.B. operators. Worst of all, someone deliberately broadcast false reports. On Sunday there was a C.B. alarm of a nine-foot wall of water coming

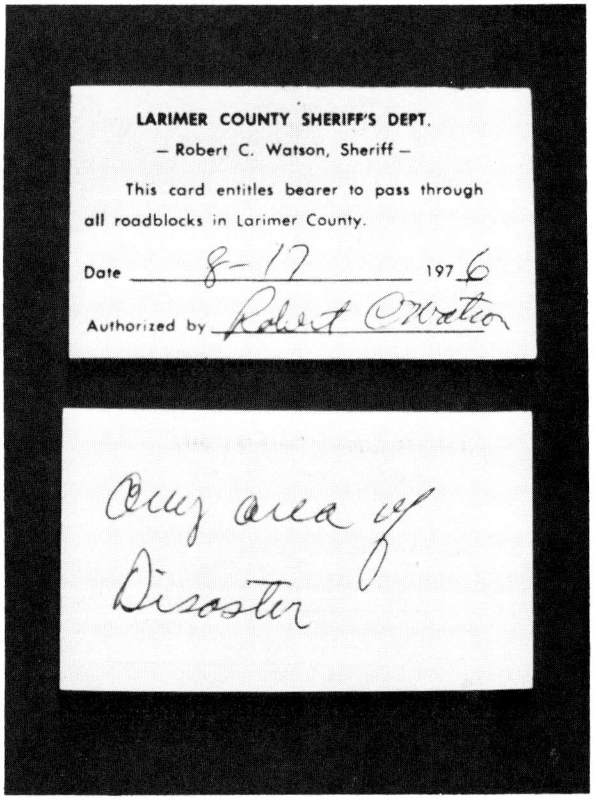

LARIMER COUNTY SHERIFF'S DEPT.

— Robert C. Watson, Sheriff —

This card entitles bearer to pass through all roadblocks in Larimer County.

Date 8-17 1976

Authorized by Robert C Watson

Only area of Disaster

down the canyon, and it included minute-by-minute updates.[45] This lie disrupted rescue operations, caused unnecessary departure of residents in low-lying areas, and engendered enormous anguish for those who had survived one calamity only to hear they had to face another. There was despicable cruelty in that false C.B. report.

The Sheriff also had a potential problem with looting. At least, he thought he had a problem, and acted to prevent it with extensive security measures. From the start, the Sheriff sealed the canyon with roadblocks and scattered patrols of seventy-five to a hundred men through the area. This was one of the reasons to use a military police National Guard unit. They were trained in law enforcement.[46]

Watson also invoked a pass system so that only authorized persons could get through the roadblocks. It was not particularly rigid, but there were arguments at the checkpoints. Some of the volunteer deputies were too zealous. James Lloyd, for example, had to use his official state license plates to escort a line of outraged residents through a roadblock so that they could care for their property. The passes and roadblocks, however, were of mixed merit. The residents became equally outraged when the Sheriff, hoping to

Police agencies often consider looting a major problem in disasters. There is some debate about the extent of looting, but this sign appeared at Drake.

To suppress looting and control sight-seers, the Sheriff set up road blocks, instituted a pass system, and ordered mounted deputies to patrol the canyon.

eliminate the barriers, lifted them and let through a swarm of gawking sightseers.[47] There was no way for the Sheriff to win on this issue.

Sociologists argue that looting is not a problem in a disaster and that police personnel can be better utilized.[48] Richard Huffsmith of Cedar Cove told about a neighbor with a torn wall which exposed some expensive cameras standing on a dresser. There was no difficulty. Yet, another neighbor, Ernest Stille, surprised seven looters in his home.[49] The police arrested only twenty-two people for looting and said that it was no great problem because it had been prevented.[50] The protective instinct of the police was strong. There is no way of judging whether or not looting was a valid concern for the Big Thompson. A few days after the flood, however, the Sheriff reduced the security patrols and encouraged the owners to return to help protect their property.[51]

Complicating the security effort was the great influx of volunteers and sightseers. It is common for outsiders, persons not involved in the disaster, to converge on a site. They are curious; they want to help; they are often unknown to those in charge; and they are a nuisance. This happened at the Big Thompson.[52] Englebert reported:

> They wanted to help; I didn't know their background. I had many, many people that wanted to just rush in there and do something. Well, I didn't want to put in any more people . . . than we already had in there. We wanted to get people out, not put them in there that first day, or even the second day. And I had more people wanting to go in than we were getting out. So, we might have ruffled a lot of feathers the first couple of days. We told people, "No, stay out!" . . . We were quite blunt with people. We would say, "No, we can't use you right now. See me in a day or two."[53]

Watson tried to send the excessive volunteers to the Salvation Army or Red Cross.[54] In one instance, he had to remove an entire group. A squad from Colorado Springs appeared wearing green shirts with epaulets, boots, badges, Stetsons, and enormous handguns. They looked and acted like police, and called themselves the "Colorado Rangers." They were not police, however, and had no authority. They were a private military organization. When he became aware of their presence on the roadblocks, Watson eased them off and sent them away.[55] He had trouble enough without them.

Also converging, but with more validity, were newspaper people from around the nation. Despite weariness, Watson held frequent conferences and accommodated the journalists as long as they did not interfere. There is no evidence that the Sheriff was anything but honest, open, and informative about what was happening. There was no deception, or coverup of errors. The news workers, for their part, cooperated, stayed out of the way, and actually helped the police by telling the public about needs and restrictions. There was little complaint from either side. Watson thus circumvented another common problem in disasters—the lack of public information.

The Sheriff and his key staff put in long intense hours without rest. They lacked back-up personnel. Englebert worked 174 hours in an eight-day period. He got his first sleep, two and a half hours, on Monday night. At first, Englebert gave verbal orders, but after three days they had to be written because he was fatigued and "fuzzy." He could not remember what had been done. They began some rotation then and, at the end of the week, reduced their numbers. On Sunday, the Sheriff closed the command post and pulled back to the regular Fort Collins headquarters.[56] He and his staff had done a commendable job. Larimer County could be proud of them.

Unfortunately, not everyone who came out of the Big Thompson Canyon was alive. The river was not kind to its victims. Most of the bodies were nude except for jewelry. They had been violently tumbled underwater against rocks and were battered, torn, and broken in hundreds of places. The abrasive, silt-laden water had removed hair and partially tanned the skin. The corpses were found in bits and pieces along the banks, in piles of debris, buried in silt, and trapped in crushed automobiles. They were scattered throughout the rivercourse from Glen Haven and Glen Comfort to the Interstate highway seven miles out on the plains, east of Loveland.[57] Fourteen-year-old Susan Wells discovered one of them at Glen Haven. "It was awful," she said. "He was stuck in a tree. All you could see was his waist up. There was junk and muck all around him."[58]

When a body was uncovered deputies would place it in a black plastic bag and take it by truck or helicopter to a Loveland funeral home. There,

The search for bodies began on Sunday. National Guardsmen scoured the Big Thompson floodplain near Loveland. *Photograph courtesy of Loveland* Reporter-Herald.

according to regulations, the corpse was cleaned and preserved before being sent to the morgue at the old hospital. The veins would not hold liquid, so the morticians sprinkled the bodies with powdered formaldehyde. This was an irritant to the people who had to work with the bodies at the morgue, but it was the law. After identification, the deputy coroners put the body into a refrigeration truck until claimed by a relative. Despite its morbidity this emergent identification system, developed by volunteers, is one of the unique and valuable features of the Big Thompson disaster.

On Sunday, the day after the flood, searchers found forty bodies, most of them beyond the Narrows on the floodplain. The next day the count rose to sixty-nine and on Tuesday to seventy-two. It reached eighty-four on August seventh and one hundred-four a week later. The death list continued to rise to one hundred nineteen on August twenty-first; one hundred thirty-four on September sixth; and one hundred thirty-nine on September twenty-sixth. This was the final count, although six people who were confirmed to be in the canyon remained missing.[59]

After the emergency phase of the first several days, the Sheriff concentrated on an exploration for the dead and people possibly buried alive. Search and rescue teams scoured the canyon. Thirty "sniffer" dogs from around the state snuffled through the rubble. The dogs, as well as the mechanical "sniffers" loaned by the U.S. Army, proved useless. There were too many confusing scents from dead fish, skunks, and other animals.[60] Body recovery, consequently, settled down to a painstaking, cooperative effort with the crews hired to remove debris. Spotters walked along and stopped work when the machines revealed anything resembling a body. There was an effort to treat victims with respect. A deputy reassured the press that operators "could peel an orange" with their road equipment. That was doubtful, but indicated that care was taken.[61]

Debris piles often contained the bodies of victims. *Photograph courtesy of Michael Charney.*

The only helicopter accident of the entire disaster occurred during this phase when a small helicopter, struggling to lift a 350-pound body, struck some rocks and crashed. No one was hurt, but the wreckage delayed road construction in the upper canyon for a brief time.[62] After the ground-level coverage of the canyon, it seemed reasonable to work with the crews removing the refuse. It was safest, cheapest, and most thorough. For those missing their loved ones, this decision, understandably, could not be satisfying. Helen Jean Wylie who had lost her parents, and James Marshal wrote from Riverside, Connecticut to the Colorado newspapers:

> This gives us the clear feeling that Colorado can find ways and means to get federal dollars and making a buck off future tourists, but they can't find the personnel or funds to go in and return our loved ones to us. . . . It seems more important to restore Highway 34 to the people of Colorado, even if it must be built upon our dead. Tourism uber alles! The dollar sign must be your state emblem. Colorado cares much for ecology, but nothing for humanity! . . . We are bitter. We are angry. We are, I think, understandably upset with a state that issues flood warnings two hours after the flood has occurred![63]

Readers replied in defense, with letters of their own, but there could be no convincing answer. It was a painful business. Despite the search there may be victims no one knew about—campers, nomads, people "on the road," drop-outs. If so, they remain buried under tons of rock and silt, or, perhaps, ground to pieces. As Captain Thomas mused, "In two years a farmer will be plowing in his beet field and will recover a skeleton. And then, we'll wonder, 'Is it a murder, or is it a flood victim?'"[64]

At first, the circumstances concerning body identification were chaotic and overwhelming. The corpses, in horrifying condition, were stacking up and no one knew how to match them to the descriptions given by anxious relatives milling about. Without discipline, the lists of missing persons ballooned in Loveland and Estes Park. Thousands of inquiries poured in from all over the country. The situation was out of control. It was brought to order by an unlikely hero—a Colorado State University anthropologist.

Every car found might contain victims. *Photograph courtesy of Fort Collins* Coloradoan.

Each wrecked car had to be searched. *Photograph courtesy of Fort Collins* Coloradoan.

Bodies were placed in black plastic bags, removed from the canyon and sent to Loveland for identification. *Photograph courtesy of Estes Park* Trail-Gazette.

A search through the piles was necessary. Dogs and mechanical "sniffers" were of little use due to the confusion of odors. *Photograph courtesy of Michael Charney.*

The river carried bodies onto the plains to this point near Interstate Highway 25. The buildings at Loveland can be seen in the background.

The abandoned Loveland Hospital became a temporary morgue when more space became necessary.

The task of identification and handling the missing list belonged officially to the county coroner. He turned the job over to two volunteer deputy coroners. The younger deputy, in his 20s, Patrick Allen, was a pathologist at McKee Medical Center in Loveland. The older deputy, near retirement age, was Michael Charney. He was a physical anthropologist at Colorado State, with forty years experience in the identification of skeletons for law enforcement agencies in Texas, Idaho, and Colorado.

Fred Fishburn, a Loveland mortician, overwhelmed by the number of bodies, called the county coroner, Robert Schmidt, at 7:00 a.m. Sunday. Schmidt called Charney at noon. The professor had not heard about the flood. His retort at the news of twenty-five bodies was, "What do you have, a mass murder?"[65]

Charney, a vigorous, pipe-smoking, bushy gray-haired man, climbed into his beat-up white Porsche sportscar and drove to the Loveland funeral home. There he found the bodies, hosed down and naked on the floor under paper sheets. He had to have more room and took over part of the old Loveland hospital. It was available and empty. There was adequate space, but only four working air conditioners and one live telephone. There were no gowns, gloves, guerneys, working tables, desks, or even paperclips. All of these had to be acquired. For the moment, he worked on the floor.

Contrary to rumor there was no disease danger from the bodies, except tetanus. Equipment, supplies, and volunteers came in. Charney and Allen borrowed a portable X-ray machine from a Loveland doctor, and refrigerated meat trucks from the Union Pacific, Timpte Trucking, and Monfort of Colorado. These were needed to control the penetrating, pungent odor which, by Tuesday, became overwhelming. They even obtained a small computer from Hewlett-Packard.

The goal of the identification work was to discover those physical traits which make a person unique—different from any other. They searched for the uncommon feature and tried to match it with that of a missing person. The Sheriff's Office

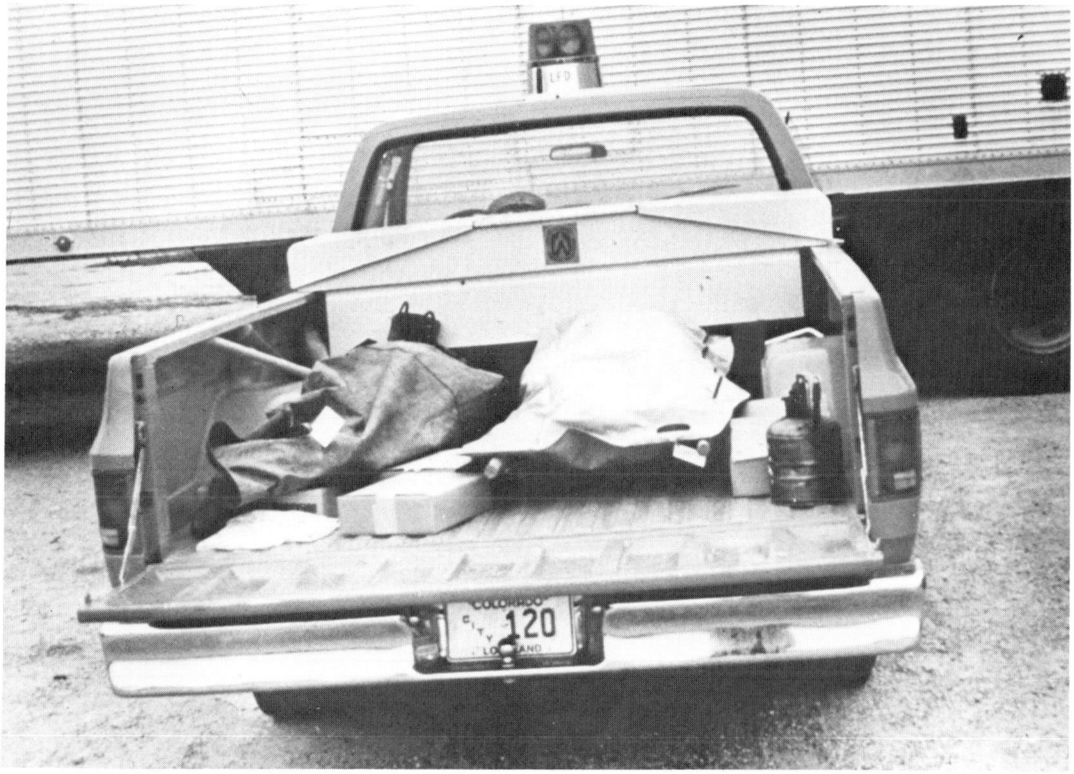

The bodies arrived in battered condition, devoid of clothing. This created a major identification problem. *Photograph courtesy of Michael Charney.*

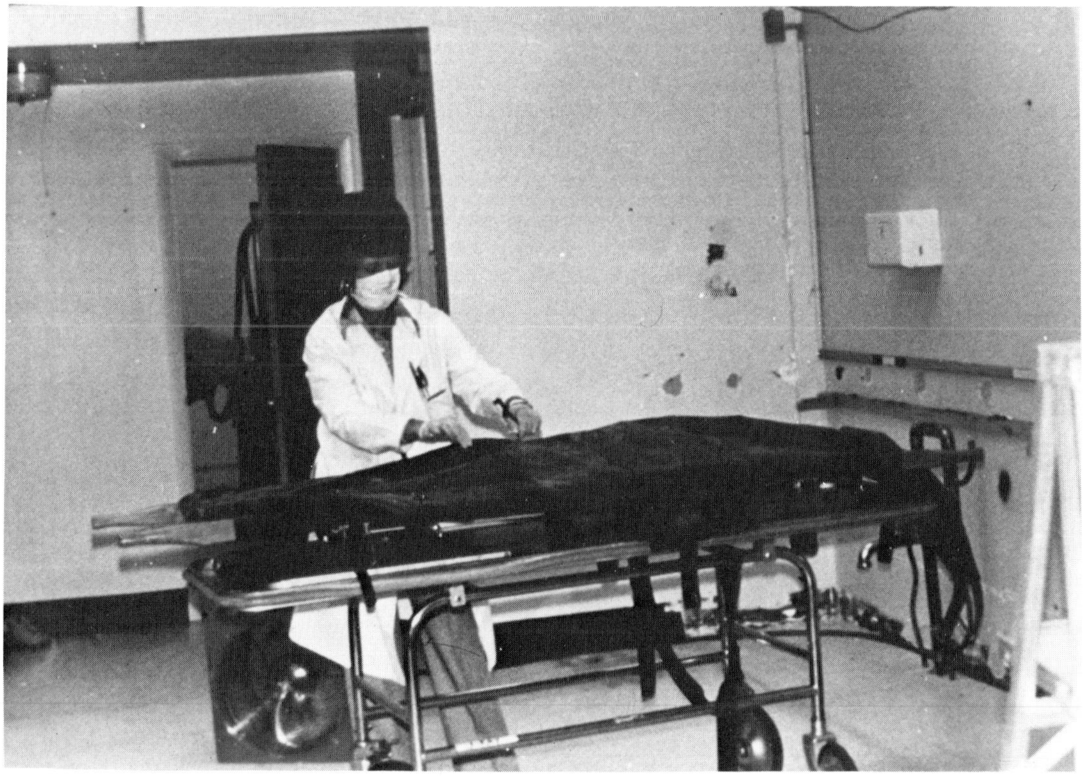

The bodies were brought in on gurneys for examination. *Photograph courtesy of Michael Charney.*

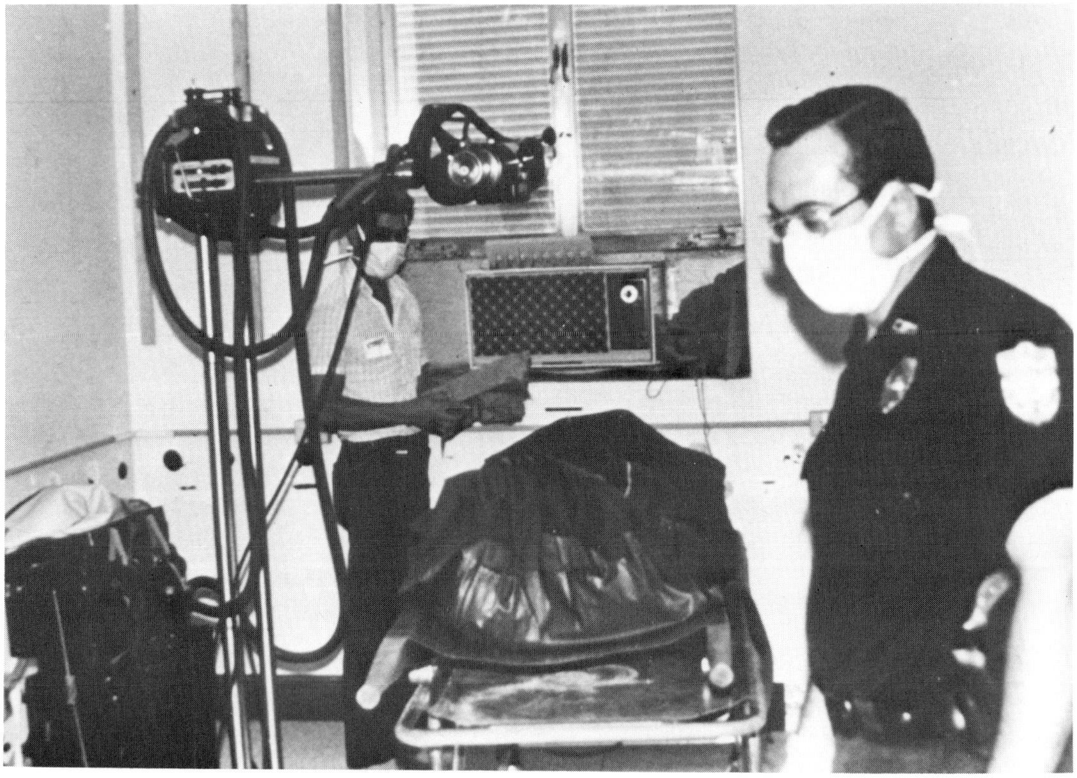

A Loveland doctor provided an X-ray machine for use in the morgue. With it, the deputy coroners were able to locate old bone breaks which aided in identification. *Photograph courtesy of Michael Charney.*

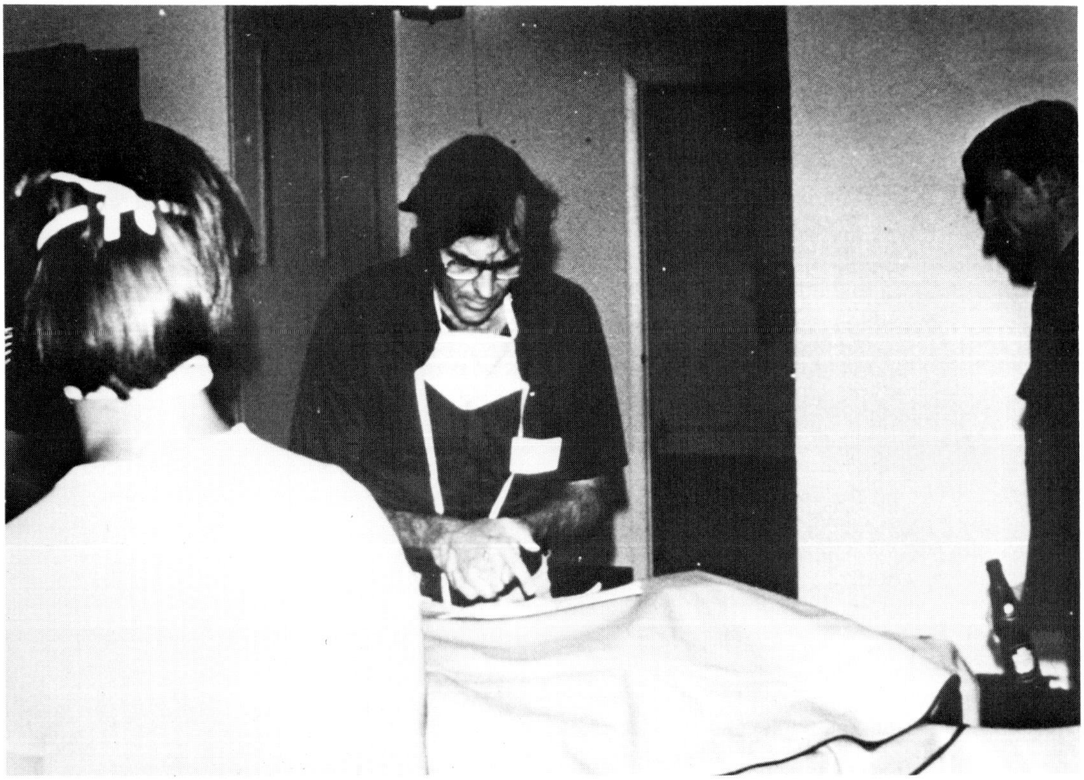

Because of deterioration, fingerprints shortly became impossible to take. Dental exams and X-ray evidence developed into important tools for body identification. *Photograph courtesy of Michael Charney.*

Following examination, the morgue workers preserved the bodies in refrigeration trucks. *Photograph courtesy of Michael Charney.*

and a visiting F.B.I. team helped with fingerprints. After five days, due to deterioration, fingerprints became impossible to take. Teeth and dental records then became the most important identification tool. There were people, however, who had not been to a dentist. When dental charts were available, they were seemingly all different. Charney and Allen needed help on this point, and they received it from an unusual source.

Prompted by Governor Lamm's office, Lt. Col. William Morlang, an Air Force forensic dentist, called Charney and offered to help. Thinking that the F.B.I. would take care of matters, the professor said that Morlang was not needed. But the F.B.I. only took fingerprints, and Charney called Morlang back. The officer came in on Wednesday and set up a tooth examination procedure with a common chart to match the varied records of home dentists. With that assistance, the Loveland dentists who had volunteered could adequately conduct a mouth inspection.

The complete identification exam took only fifteen to twenty minutes, but it covered fifty-one items including height, weight, sex, age, hair and eye color, and scars. Interestingly, only a few of the male victims were circumsized, and that became an identifying characteristic. It convinced one woman that a particular dead male was indeed her former husband. Nail polish in one case and an unusual shaped ear became deciding factors in another. The X-ray machine revealed a pellet in a hand, a hip-joint pin, and a healed leg fracture in various victims. Charney and Allen assiduously sought out that special, extraordinary feature which made people physically different. They were successful; they identified all of the victims.[66]

It should be added that Charney, Allen, and the staff at the morgue worked with no promise of pay. They did not have to do this work. Yet they were professionals and thought they should contribute their expertise. They were careful, humane, and sensitive. They took pride in what they accomplished. As Charney explained,

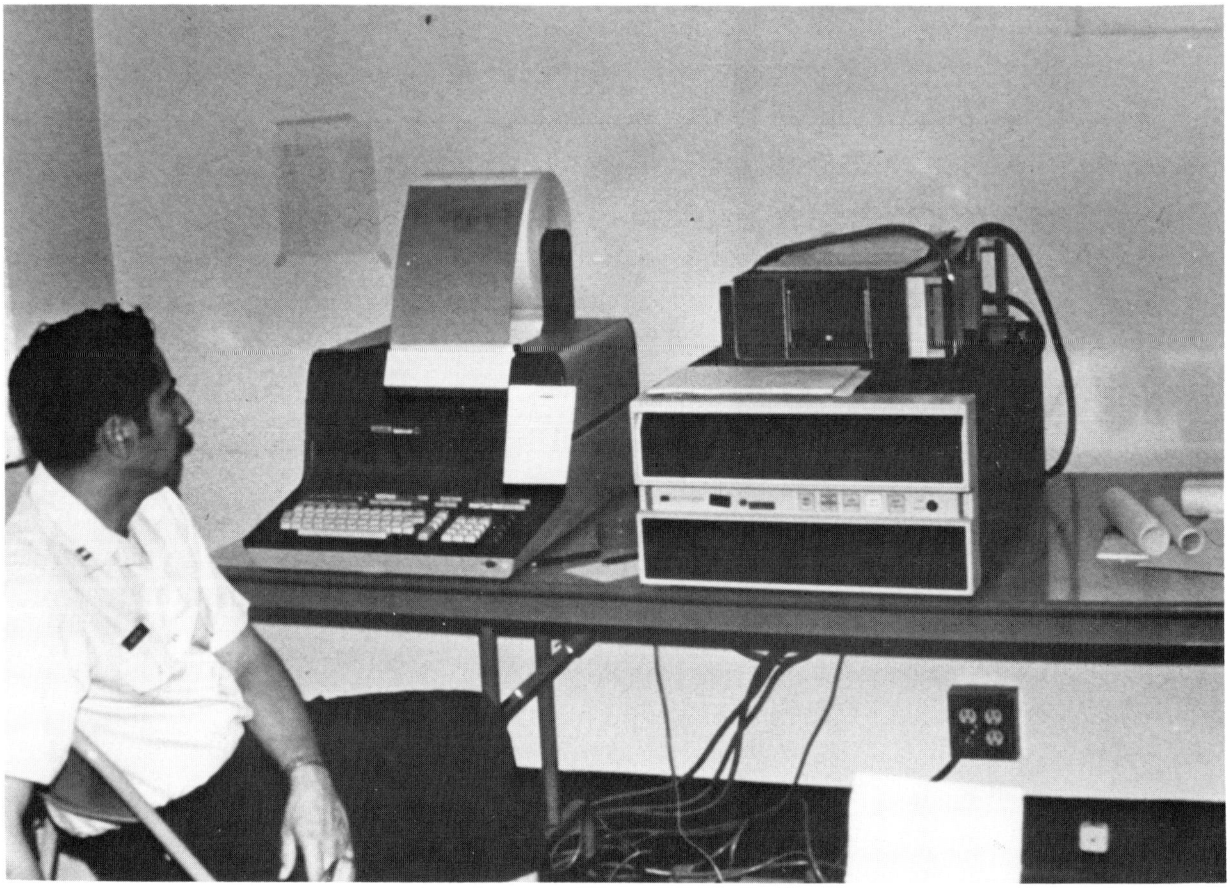

Hewlett-Packard provided a small computer to aid in the identification procedure. *Photograph courtesy of Michael Charney.*

"People had to train for this work. You do it whether or not you want—I'm making a living at the university. I've got a salary. I didn't get any all summer, but we live."[67] Colorado should cherish such people. They ennoble the human race.

Their work took place behind closed doors. There was no public viewing of the bodies, no lines of weeping women dressed in black filing past rows of corpses—a scene often connected with disasters. Charney perceived such action as unnecessary and conducive to mass hysteria. It lacked decency, and so the staff designed a different method. It involved the missing persons list and the information gathered from relatives for a match-up.

Ten members of the Larimer County Mental Health Clinic traveled to the Red Cross center on Sunday to see if they could be of service. Wandering through the crowd at the gymnasium they found that most people were well controlled and only wanted information about loved ones. They aided the Red Cross in screening refugees as they came in, discovering an alcoholic and a case of measles. By and large there was no use for them; people possessed a good state of mental health. For that matter, as it turned out, there were few long-term needs either.[68] Seven members of the mental health group, along with some county social services personnel, drifted over to the morgue where they seemed to be needed.

They scrounged for supplies, interviewed people about missing relatives, devised and revised a description sheet, and put data on the Hewlett-Packard computer. With the help of Ann Davis, a volunteer nurse, who had knowledge of hospital records, they developed a flow of information to support the identification process.

They also attacked the missing persons list. Within forty-eight hours the list was 1,000 names long. In the thin hope that there might be news, pathetic parents called long distance to report missing sons and daughters who had disappeared years earlier. There was duplication and misspelling. Mrs. Nettie Eder, for example, was really Mrs. Nettie Oler.[69] Some people reported

The list of missing persons reached over 1,000 names within forty-eight hours. People from all over the country called into the morgue searching for relatives missing years before. Women from the Retired Service Volunteer Program spent 200 hours tracking and verifying information.

themselves as missing. At Glen Haven, a man in a black cowboy hat pulled up to the roadblock and drawled with a country accent, "I'm not dead, but my wife thinks I'm missing, and I'd like to keep it that way for a few more days. But I don't want her to start settling my estate yet, so if you see her, just hedge a bit."[70]

There were also four teenagers who came in to take their own names off the list. It was embarrassing. They were supposed to be in the canyon. Their parents thought that was where they were, but they had slipped off to Denver instead.[71] It is the sort of situation that makes a parent weep for joy and stamp in anger at the same time.

The list had to be checked. Two women, Vi Dardarian and Matilda Hartung of the Retired Service Volunteer Program (R.S.V.P.) in Loveland, spent 200 hours on the phone tracking and verifying information. In four days they cut the list in half. By September 12, they had pared it to sixteen.[72] A remarkable effort.

When the information from relatives matched the data of the body examination, the staff notified Charney or Allen. They confirmed the match and contacted the relatives. "It isn't easy to tell someone that you have just positively identified a body as that of a loved one," observed Charney, "but it removes doubt and frustration and clears the mind for acceptance of their loss."[73] Not knowing what happened is the hardest to bear.[74]

Painful to see were those awaiting news. Phyllis Conley, the wife of the Estes Park policeman who sacrificed his life at Waltonia, for example, waited. "I think it's just a matter of time," she said. "Unless circumstances completely overwhelmed him, he's up there. I won't leave him up there. I just couldn't do that to him."[75] Members of the Mares family also waited. James Mares lived at Cedar Cove with his family and was talking to his mother in Denver when the phone went dead. The house was swept away and a neighbor

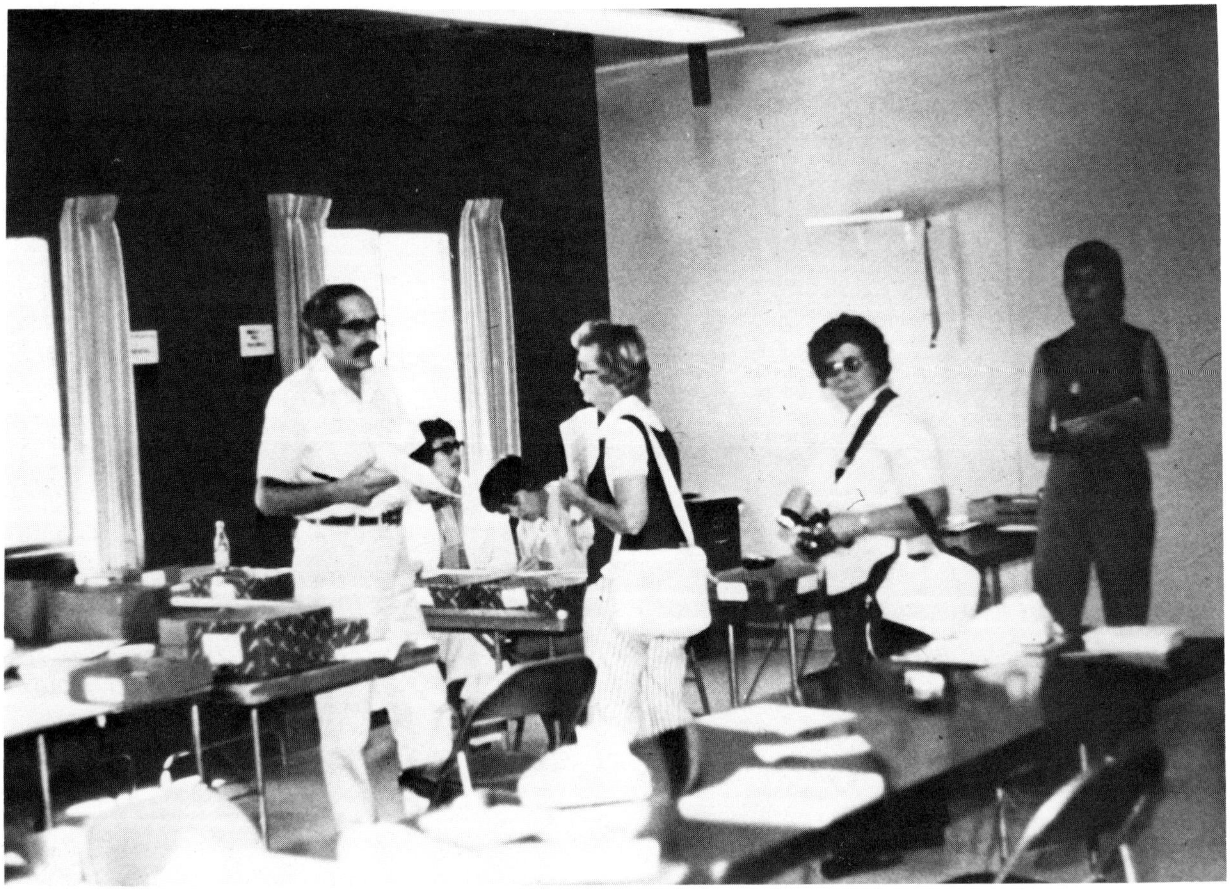

Volunteers matched data about missing persons with evidence from body examination. Michael Charney, deputy coroner, is shown talking to one of the volunteers. *Photograph courtesy of Michael Charney.*

saw the family's four-wheel drive International Scout, lights on, sinking in the water. McCoy Mares walked the Big Thompson four times searching and slipping in the mud and rocks for his brother. He found other bodies and the family bloodhound still chained and drowned, but he could not find the family. He also waited.[76] Such people deserved help and accurate information, humanely handled.

When the deputy coroners met with the family often a minister was there to ease the pain. Father Roger Mollison, for instance, would say mass and pray. Even then there was a feeling of helplessness. "About all you can do is just be there to talk to them, and let them know you care," he said.[77] Some people wanted to see the body, but this was discouraged because of the broken condition and deterioration. Sometimes they showed a photograph of the head, and, of course, if a relative insisted, they allowed viewing. Most people chose not to do so, and accepted Charney's explanations.[78]

After eleven days the Mental Health contingent pulled out, but the morgue continued to operate until mid-September.[79] By then the special facilities were no longer necessary. The unpleasant task was over, and the professor returned to his classes at Colorado State University.

After identification and notification, relatives could claim the bodies and move them for burial. *Photograph courtesy of Fort Collins* Coloradoan.

Governor Richard D. Lamm conferred with disaster officials and flew through the canyon by helicopter.
Photograph courtesy of Loveland Reporter-Herald.

Chapter IV

Recovery: Pointers and Scoopers

The final stage of a disaster is the recovery of the people and the land. In physical appearance, except for man-made structures, even in the acute observation of the Park Rangers, there was no great difference.[1] There were, of course, some new washes, deeper gullies, and scouring in the river bed. The fish were gone, probably washed into Nebraska, but they could be replaced through the state fish hatcheries. There was also loss of vegetation and, as a result, the U.S. Forest Service planted grass seed on the mountain slopes in the Fall of 1977. They sprayed over the seeds a mixture of wood mulch and fertilizer to hold and encourage growth. The following spring the Forest Service recruited 250 volunteers and planted 4,000 trees and shrubs.[2] By giving nature a boost, the foresters masked the scars of the catastrophe. With the debris cleaned away, a family touring the canyon in the summer of 1978 would have difficulty noticing that anything had happened unless they knew specifically where to look.

The people, however, did not recover with such ease. This long-running two-year phase was thoroughly infused with legal regulations and unresolved ethical questions that could find settlement only in the political arena. The basic points of dispute involved the extent of aid properly due to the victims and the action necessary to prevent a repetition of the disaster. Is the government responsible for people who choose to live in harm's way? Should a government, or society, act to mitigate future disasters at the expense of reduced individual rights? These questions at the Big Thompson, fundamentally, represent a continuation of that most profound inquiry of American political history—what is the correct balance between the liberty of the individual and the good order of society? It is a debate that reaches back to the founding of the Republic and the writing of the Constitution.

The politicians gathered early at the scene of the disaster and acted quickly to release current aid monies. Senator Gary Hart and Representative Jim Johnson, for example, visited the Big Thompson and then returned to Washington, D.C. to help from that point.[3] Hart heard Governor Lamm speak about the catastrophe at the noonday Centennial celebration in Denver. The senator then flew over the canyon—most visible from his altitude of 700 feet was the road destruction—and sent his staff members to Loveland to see what was needed. In Washington, he researched pertinent legislation, joined in the request for the official declaration of a disaster area, and helped obtain food stamps for the survivors. When asked about governmental responsibility in such a situation, Hart responded:

> I wasn't elected God, you know, I was elected to do what I can, and what I'm going to try to do is get as much for those people as I can. . . . Nevertheless, I find it rather ironic that the people who are often quite outspoken about their independence and their distrust for the federal government are, under circumstances like that . . . often very inclined to turn to the federal government for help.[4]

State Representative James Lloyd of Loveland was the first politician at the disaster site. He conferred with the governor on the night of the flood and later cut through bureaucratic regulations to ease the burdens of the people. *Photograph courtesy of Loveland* Reporter-Herald.

Governor Richard D. Lamm toured the area on Sunday morning His work was important for the long range recovery of the canyon. *Photograph courtesy of Loveland* Reporter-Herald.

Representative James Howard, a New Jersey Democrat vacationing in Estes Park, also flew over the Big Thompson. It was his committee in the House of Representatives that promoted the Disaster Relief Act of 1974. This law provided money to replace highways; it was an important enactment in regard to Big Thompson recovery.[5] Of the various politicians, Governor Richard Lamm was the most important. After whirling through the canyon at dawn on the morning after the storm with Sheriff Watson, he returned to Denver to work on a message to President Gerald Ford. Amidst the festivities of the Centennial, his staff put the message together and it was sent off at 7:30 p.m. Ford replied at 9:00 a.m. the next day, Monday, and declared Larimer County a disaster area with access to federal aid programs.[6]

Beyond that, Lamm issued four special executive orders on his authority under the state emergency act. He waived highway regulations which called for hearings where private property was involved. This allowed the rapid construction of a pioneer road through the canyon. It could not follow the old route precisely because of the destruction. Second, he permitted trespass onto private property for the restoration of public services. Third, he ordered the removal of debris, and fourth, he gave authority to Larimer County to regulate building in the Big Thompson for six months.[7] In addition, Lamm participated in the formation of a complex funding package for the purchase of land within the floodplain.

On a somewhat more personal basis, State Representative James Lloyd also figures prominently. He was at the Merri-Ax when the evacuation operation started. He kept the governor informed, obtained radios from the Forest Service, found blankets through Civil Defense, and cut some red tape at the driver's license bureau. The bureau was going to make all persons who had lost their license take the Colorado driver's test. Lloyd persuaded them simply to issue a temporary permit without a test. He also sent twenty Job Corpsmen who wanted to assist into the Glen Haven area. He found some free culverts at the Pueblo army depot and the Job Corps went to work cleaning up and repairing roads. Lloyd

The Big Thompson was cleared of debris. *Photograph courtesy of Estes Park* Trail-Gazette.

sought permission for this free work from the Larimer County Commissioners, but they procrastinated. The corpsmen, meanwhile, went in, worked for five weeks, and left. At the end of the time the commissioners still had not decided whether or not to use them.[8]

The commissioners had problems, among them an election in November. Meanwhile, there existed a whole panoply of organizations and programs to help the survivors. To efficiently provide information and service, the Federal Disaster Assistance Administration set up "one-stop" centers in Loveland, Estes Park, and Fort Collins. Flood survivors needing help first came in and checked with the Red Cross to see if they were on the missing list. Then they were told what aid was available—food stamps, housing, loans, unemployment benefits, social security, veteran's assistance, travel money—and how to apply for help. Representatives of the various programs were there to answer questions and aid people in filling out the various forms. When they were through, visitors were given an exit interview to be certain they were satisfied. Although for most programs people had sixty days in which to apply, 941 persons visited the centers during the

A workman assigned to watch for bodies worked with road machinery.

eight days they were open. Most of the applications for the various programs came during this eight-day period. After the centers closed, people had to travel around to individual offices.[9]

The Department of Housing and Urban Development moved thirty-five mobile homes to Loveland from Dempsey, Texas for temporary shelter. With these and other facilities, they housed 165 families. H.U.D. later financed a $1,421,000 project to construct 72 units through the Loveland Housing Authority. Big Thompson victims had first choice. The Colorado Housing Finance Authority, in addition, made available $300,000 for 7-1/4 percent, 30-year, 5-percent down loans for survivors with less than $14,300 yearly income. This Colorado contribution, however, came a year and a half later.[10] Its restrictions and lack of timeliness made it of questionable value.

People also had to wait for food stamps. In this instance it was only for three days. That was the rule. A person could get pretty hungry in three days. At the request of the Colorado senators, Secretary of Agriculture Earl Butz waived the waiting period. In another reversal, the department allowed aid to twelve families which had adequate, but inaccessible, personal resources. The food stamp program lasted thirty days and 160 families received free stamps. The cost was $8,242.[11]

The Small Business Administration also reversed itself and permitted the financing of second homes. It offered 30-year loans at 6-5/8 percent interest for houses and businesses. They received 71 applications.[12] The problem for victims was, of course, that any prior loan continued, regardless of the fact that the property had been destroyed. Only one person in the entire canyon had flood insurance. The S.B.A. interest rate, therefore, became a point of irritation for the residents of the Big Thompson. They wanted loans that carried little or no interest charges.

There were slightly over two hundred applications for both Family Assistance Grants and unemployment payments. Survivors also benefitted from the work of the Red Cross and Salvation Army.[13] Residents received some free public services as well. These were a mixed blessing. A

Contractors hauled the debris from the canyon to a landfill near Loveland.

combined federal, state, and local health team surveyed the canyon to test the sanitation of wells, locate animal carcasses, mark buildings for destruction, and help supervise debris removal.[14]

The U.S. Army Corps of Engineers directed the extraction of the debris and the Gypsum Mining Company of Chicago provided a dump site near the mouth of the canyon. It had some pits to be filled. The Corps contracted with various companies for the effort and, for the most part, used local workers and equipment. The debris was thus removed from the lower end of the canyon. Wrecked cars were hauled to Estes Park where they could be claimed by owners and inspected by insurance agents.[15]

Although contractors had the authority to enter private land, they still sought permission from owners. If the owners wished, they could clear their own property, pile the debris by the road, and the workers would haul the refuse away at no charge. The Health Department condemned houses that were more than fifty percent damaged and marked them with a large "X" for the bulldozers. The owners could challenge the "X" rating if they wished, but there was little choice. The buildings were too far gone to be repaired.[16]

Mildred Crockett who lived just west of Loveland agreed to let the Corps take her home. A private contractor had told her it was not worth saving. The foundation had eroded and the main wall was crooked. All that could be salvaged were the windows. Yet, Mrs. Crockett had lived there for thirty-five years. She reared her children in the house and it was filled with memories of her husband who had died in 1974. "My heart just pounds and pounds and pounds," she lamented. "It's telling me I'm losing my home, I guess . . . but I'm thankful we still have our lives." She maintained a stout spirit and praised the workers for efficiently removing the fifteen-foot-high piles of debris without harming her cottonwood trees.[17]

Grace Srb of Lincoln, Nebraska, did not see it that way and sued the State of Colorado, the Department of Health, and Governor Lamm for $83,000 after her home at Drake was removed without permission. The case remains unsettled

Buildings damaged less than fifty percent could be restored according to floodplain zoning laws.

within the courts.[18] This was the only instance of that nature and attests to the reasonableness of the condemnation procedure.

It took two months to clear the canyon once the Corps of Engineers awarded the first contract. It cost $1,041,000 to remove 320,450 cubic yards of debris.[19] Larimer County provided free temporary bridges—culverts covered with dirt—and tax reductions for damaged property. The R2-J school district absorbed a cost of $3,280 for the operation of the Red Cross center at the high school during the disaster.[20] Electric power was restored throughout the canyon in three weeks, and something can even be said in praise of the often maligned telephone system.[21]

Mountain Bell worked around the clock to restore service. They flew a microwave unit from Arizona to give emergency long-distance access from Estes Park, and the company worked quickly to provide extra lines at the high school and the Sheriff's command post. Normal peak load for long distance from Loveland was 1,500 calls per hour. On Monday after the flood it was 14,000 and on Tuesday, 11,000. The capacity of the equipment was sufficient; there was no overload. Lines were down, but it was unreasonable to erect poles and string wire when no one was certain about the route of the restored highway. The telephone repairmen laid the wire on the ground and in eight days service was back to normal throughout the area.[22]

The phone company possesses a special spirit when it comes to emergency work. Its crews are prepared for such events and work diligently to repair facilities and extend aid in communications. Rick Hays, a Mountain Bell manager from Greeley, for example, talked himself hoarse relaying free emergency messages through the microwave unit he set up at Estes Park.[23] At the Big Thompson, the dedication of the company inspired stories. This one cannot be substantiated, but, according to hearsay, when Watson's motorized expedition reached the Waltonia area on Monday on a mission to rescue stranded motorists, all that was found was a Mountain Bell truck. It was parked beside the road with a regu-

The Mennonite Disaster Service provided volunteer labor for clean-up purposes throughout the canyon. This scene was at Glen Haven. *Photograph courtesy of Estes Park* Trail-Gazette.

lation red warning cone placed neatly alongside the rear wheel to warn passing motorists. The driver, tools dangling from his belt, was at the top of a nearby pole working on the lines.

There was also the unofficial aid for the survivors. Rich White, a Loveland businessman who once worked with Danny Thomas to collect money for St. Jude's Children's Research Hospital, gathered $240,000 for the Loveland Chamber of Commerce. It came from a variety of sources; the Colorado Springs Jaycees sent $1,700; Statitrol Corporation gave $10,000; a C.B. operator called "Coin Hustler" collected and contributed $1,100. The purpose was to fill in the "gaps" left by other support programs.[24] There were benefit softball games, basketball contests, dances, and even a bike race.[25] The employees of the Rocky Flats Nuclear Plant gave their first safety bonus, $6,050, to the cause, and Charles Sanders of the Horsetooth Inn in Fort Collins sold a recording for the benefit of victims.[26] There is no evidence of any misuse of these informal funds, but it would be interesting to examine how they were spent and the criteria for distribution.

Two organizations, the Mennonite Disaster

Irma Sidwell, Postmistress at Glen Haven, supervised the clean-up of the Post Office. *Photograph by Estes Park* Trail-Gazette.

Service and the Interfaith Task Force, supplied immediate direct help. Both performed similar work—clean up and repair. The main difference was that Interfaith stayed longer and provided somewhat more comprehensive aid. To a resident standing in mud a foot deep, no one presented a more welcome sight than these volunteers.

On the morning after the flood, there was a request at the Brethren in Christ church service in Abilene, Kansas, to send the Mennonite Disaster Service to the Big Thompson. This organization originated in Kansas in 1951 to deal with the devastation of floods and hurricanes. Since that time it had spread nationwide through the various branches of the Mennonite Church. The M.D.S. asked Sam Minter, a wise, white-haired, retired farmer to lead the group. He traveled to Loveland, stayed first at Campion Academy, and then moved to the Sylvan Dale Ranch which became the M.D.S. headquarters.

After the news spread through the church network, volunteers poured in from Kansas, Nebraska, Iowa, South Dakota, and Colorado. They included carpenters, masons, plumbers, electricians, and unskilled helpers. Their motivation was to give witness to their faith through good deeds. They came to work, not to preach, and people who needed help had only to ask. Minter, upon request from a resident, would send out a crew. He kept about fifty people working all the time.[27]

"Someone could have handed me a $100 bill and it wouldn't have meant as much as seeing those big, strong-shouldered Mennonite men with shovels coming to help us," said a lady at Drake faced with five feet of mud in a basement.[28] In another instance as the men began work, a large, black bear came snuffling around the corner of a garage to see what was happening. The workmen, who had just arrived from Kansas, in turn, lined up to look at the bear. As they eyed one another with mutual curiosity, one of the Mennonites said, "He must be hungry," and took out an apple. He rolled it to the bear who scooped it up, waded the stream, sat down on a flat rock, and ate it. The bear then waddled back, accepted some bread, and padded off downstream.[29] Feeding the hungry is also in the Mennonite tradition.

Twenty M.D.S. volunteers worked at Glen Haven and stayed at the Inn. Bill Wells, the owner, had remained behind to protect his property, and it was a good idea. All the experts came tramping through. The rescuers assumed that since the town had been destroyed, there was no need to protect anything. They propped their muddy feet on the furniture, and more or less took over the building as a headquarters. Wells, nonetheless, cooperated and gave his food away. It was spoiling, so he made a "$100 stew" for the National Guard. The army never lived so well. The stew contained Chateaubriand cuts of beef, breasts of chicken, cherry tomatoes, and ten pounds of mushrooms. For dessert, the soldiers were able to choose from fudge-nut, cheese, or German chocolate pie.[30]

The Mennonites were welcome. They scrubbed and cleaned; they did the dirty jobs without complaint. Ted Hanes who was at the Inn with Wells explained, "They came to work and that's just what they did." The various officials and rescuers, Hanes observed, were "more pointers than they were scoopers."[31]

The Interfaith Task Force performed much the same sort of service as the M.D.S., except to a greater extent. Interfaith coordinated church efforts, distributed donations, and solicited help from unions and building suppliers. There is no permanent national structure for this organization. Rather, it rises like the legendary phoenix from the rubble of destruction. The idea passes from one disaster to the next, carried by people who have been involved in other Interfaith efforts. James and Donna Hart traveled from Xenia, Ohio to Loveland to advise about organization and help select the local leader, Reverend Robert Schelling. They transported the idea and brought it to life. This was the ninth Interfaith Task Force in the nation.

Robert Schelling had grown up in Loveland and worked in Boulder. He had just completed seminary work and was searching for a church. When the Episcopal priests of Loveland and Estes Park requested help, their bishop sent Schelling. At an organizational meeting, with the help of the Harts, the people selected him to head the Interfaith work. He proceeded to obtain office space, volunteers, equipment, money, and supplies. He coordinated all of this, operated with administrative costs of about $50 per week, and held to the task for sixteen months.[32]

The Task Force wanted to help people like the old man who said, "We came out of there nonpersons—no address, no telephone—temporarily, you almost lose your identity—you're in a state of suspended animation."[33] Interfaith gave them something tangible to hold onto. One resident of the canyon watched eight Interfaith electricians wire a rebuilt wing while, at the same moment, a bulldozer tore away a damaged portion of his home. The owner was laughing and crying at the same time.[34] Neither Interfaith nor the Mennonite Disaster Service pressed a religious opinion, but if someone wanted to talk about faith, they would respond. Religion and a feeling of memorial, however, were not neglected. It was a part of the healing process.

Pope Paul VI expressed concern and sent a cablegram through the Denver archbishop. "His Holiness prays for those who have died and conveys his condolences to the bereaved families," it read.[35] The Campus Crusade for Christ ran a combined memorial and advertisement in the newspapers which said, in part, "It is certain that these seven women who were part of the staff of Campus Crusade for Christ International will be deeply missed by family, friends and fellow staff members. But what those women and their families would want you to know is that they are more alive than ever—in heaven."[36]

In mid-August a memorial service was held at the Loveland Community Building. Five hundred people gathered to hear Reverend Philip Dunford of Westminster United Presbyterian Church in Fort Collins say, "Someday, the Big Thompson Canyon scar, filled with our hurt, pain and suffering will grow back. . . . That pain and that hurt is human. The 'human' was created by God, just like the canyon was. No one can take away the tragedy and all the hurt, and all the pain, but all of us can think about something higher." The service ended with the gathering singing the old Universalist hymn "Amazing Grace."[37]

It was comforting to pause, pray for the dead, and remember the past. In Estes Park, however, there was a deep concern for the future. It was a tourist town and its most important business month was August. Forty percent of the annual income was earned then. It was the time to pay bills, to make it for the year. On a normal summer day 9,000 cars with 27,000 people came to Estes Park over U.S. 34. In August, 1976, the town thus faced a catastrophe of another sort—an economic disaster. There was an impression that Estes Park had been obliterated by the storm. The first news flashes indicated that to be true. Because the telephone lines were down, the phone company stopped incoming calls with a recording which said that Estes Park was "unreachable."[38] This seemed to confirm the false information.

Corky Rogers decided that something had to be done to save Estes Park. He was the manager of the Estes Park Chamber of Commerce. His job was to promote the business health of the community. On the night of the storm he had been stopped by a mudslide on highway 36. He helped the people in front who had been hit by the mud to get out of their car, and notified the police. The police wanted him to open the Chamber of Commerce building for refugees and to help find spare motel rooms. They helped him and his pregnant wife across the blocked highway and on to Estes Park. He opened the building, found sixty-five rooms, and in the process, recognized the economic danger to his town.

During his second year of college his father died. There was no insurance and four children still at home. His duty was clear. Rogers flew home, comforted his mother, arranged for the funeral, and went looking for a job. There was a time for mourning, but life had to go on. It was the same for Estes Park. After the initial stage, it was not the job of the Chamber of Commerce to care for the dead and injured. Others were doing that. The Chamber had to care for Estes Park.[39] "We've given clothes. We've given food. We've given money and whatever we could to help the disaster victims," he said a few days later. "But there's a time now that we have to seriously consider the ramifications of the drop off in tourism, since it is our only industry here."[40]

On Sunday, while Englebert mounted his airborne evacuation, Rogers planned a campaign to counteract the loss of tourists. He called Colorado radio stations and asked them to give correct information about the town. He held news conferences, and asked the local business people for advertising money. Through the Denver agency of Frye and Sills, he set up a $30,000 campaign. They placed a message in newspapers from the mayor deploring the tragedy, but stating that Estes Park was all right. They bought time for radio announcements, and the Chamber sent out more than 5,000 news releases. They aimed at their major source of tourists—Chicago, Dallas, Kansas City, Des Moines, St. Louis, Fort Worth, Denver, and Cheyenne.[41]

Governor Lamm carried out a previously planned family vacation in the park and pronounced the town "alive and well." He also noted the potential economic tragedy if the tourists stayed away.[42] Business was sixty-five percent of normal and the barometer of the region—visitor numbers at Rocky Mountain National Park—was down by half.[43] The Highway Department erected detour signs around the Big Thompson and marked an alternate route to Estes Park over U.S. 36.[44] The state travel offices sent out messages stating the flood damage did not go beyond the canyon. A toll free phone gave the information that all was open and available in Estes Park.[45]

Rogers won. On the ninth day after the flood he noticed he had to wait in line for lunch. He was jubilant. "They're coming back!" he said.[46] Although visits to Rocky Mountain National Park were down by twelve percent, retail sales at Estes Park were up nineteen percent over the prior year. They were not as high as they might have been without the flood, but nonetheless, Estes Park suffered no economic disaster.[47]

The president of the Chamber of Commerce, Robert Mobus, worried about an accusation of insensitivity, even though he recognized the necessity of the campaign.[48] A "quickie" account of the flood published within twelve days raised the issue of exploitation and callousness.[49] The same feeling surfaced in letters to the Estes Park *Trail-Gazette.* William P. McNamara wrote:

> Just why did the Chamber people move to the

Governor Richard D. Lamm (second from right) made a public statement assuring the public that Estes Park was "alive and well." Estes Park faced economic devastation following the loss of tourism because of the flood. *Photograph courtesy of Estes Park* Trail-Gazette.

> mountains? To enjoy God's works? No, to make a buck. . . . If all we can think about is the Quantity of Dollars without consideration of the Quality of Life, then why are we here instead of Denver, or elsewhere, where the dollar reigns supreme?[50]

Another writer, Jack Dreibus, expressed the same sentiment:

> While people in our area—yea, across the nation—are still trying to recover from the shock of the loss of their loved ones and their property, still unable to fully comprehend the extent of the tragedy that has overcome them, Mr. Rogers is wasting no time in drumming up funds for our Chamber of Commerce to advertise that it is "business as usual" in Estes Park. What a classic of ill timing, Mr. Rogers, one that surely must have left a bad taste with many citizens of our community.[51]

Mobus replied with the only answer available—the Chamber and the people connected with it had done all they could for the victims, and it was not until afterwards that the Chamber moved to rescue Estes Park.[52] Was the advertising campaign callous? Insensitive? Necessary? Would the natural tendency of convergence have taken care of the economic problem? There is no way to precisely calculate an answer.

Loveland, the town at the other end of the canyon, however, concentrated on the disaster and did not become involved in a national campaign for tourists. There may have been some spillover, and Rogers stated, "I think we've done more for Loveland than Loveland has done for Loveland."[53] Be that as it may, Loveland had a more diverse economy and was not isolated in the mountains. The merchants, particularly those on U.S. 34 leading to the devastated canyon, suffered. Business declined fifty to seventy percent in August along the highway, and a Conoco gas station operator commented, "It's like a boneyard out here, and we used to need a stoplight to just

Quick restoration of U.S. 34 became an important issue during the recovery period. It was necessary for tourism.

get across the road."[54] The west end merchants advertised in the Loveland newspaper and persuaded the governor to order a new traffic sign. Originally, the Highway Department put up two signs, one to divert traffic to the south, and the other to warn, "U.S. 34 open only 7 miles west." The merchants' protest brought a third sign which read, "U.S. 34. All services available. Gas. Lodging. Food. Camping. Next 7 miles."[55]

Loveland had other problems, in addition to the loss of tourists. The flood not only leveled the hydroelectric plant, but also disrupted the water supply system. The electric plant furnished only five percent of the power; the town could get along without it. Loveland took its water from the Big Thompson, however, and there were difficulties with that. The intake gates were blocked with silt and debris; a broken sewer line in Estes Park had briefly poured raw sewage into the stream; a major thirty-six-inch water main had broken for 100 feet; and the turbidity level of the river remained high for weeks afterwards.[56]

The city repaired the broken main and cleared the water intake gates. Because of lasting turbidity, the water filters required backwashing every three to four hours. Before the flood, once every twenty-four hours was normal. The city also added extra chlorine and restricted lawn watering to every fourth day. The lawns survived and so did the people. It was not necessary to truck in water as once feared.[57] Loveland was even able to hold its county fair on schedule despite the mud which had to be removed from the site. In 1951, due to the same problem, the fair was cancelled.[58]

Loveland suffered $2,100,000 in damages to city facilities, much of this in the mountain park and hydroelectric plant. They could be repaired with federal funds, but two years later they were still unrestored. The plant was scheduled for rebuilding in Fall, 1979. At the time of the flood, retail sales had been racing along at thirty-two percent above the previous year.[59] The town ended up only nineteen percent ahead; the Big

The state highway department quickly opened a pioneer road through the canyon.

Thompson took its toll.[60] Compared to Estes Park, there was no difference. Was Rogers' advertising campaign necessary? There is too great a variation in the towns to say for certain, but the advertising probably did not have to be as extensive as it was.

Although the two towns might disagree about the propriety of advertising, they concurred completely about the immediate restoration of U.S. 34. From the outset, people wanted to know about the road. Jack Kinstlinger, the director of the Colorado Department of Highways, replied that it would be rebuilt and improved, and "pretty much where it is now" in nine to twelve months.[61] That was reassuring, but the highway was an anxious issue for people.

Residents depended upon the road for their business, travel, and communication. A change would require an adjustment, and, moreover, they might lose money. The highway, however, was more than that. It was the binding cord of the community, a means of measurement, and a point of orientation. To move or change it threatened the community order. Most residents wanted the highway right where it had been. Given the number of people and emotions involved, it would have been impossible to do anything other than reconstruct U.S. 34 in its old location. Kinstlinger predicted correctly, but there were some nervous moments.

The Highway Department went to work right away to clear a "pioneer road." This was a single-lane, gravel track for road equipment and four-wheel drive vehicles. It provided crude access for property owners, police, and clean-up workers. The engineers had to use dynamite to remove obstructing boulders as large as "elephants," and they had to scrape the river bottom in search of fill dirt.[62] It was an added insult to the canyon, as if humanity had to have the last word in scouring the canyon floor. The highway was nearly complete by the end of August, open to the public in October, and temporarily paved by the end of November. When the Sheriff lifted the barriers

Bulldozers scraped the bed of the river to find fill dirt. *Photograph courtesy of Fort Collins* Coloradoan.

on October 26, some 2,500 to 3,000 cars bounced through with sightseers taking pictures. A resident grumbled that he felt "like an animal in a zoo," but the road was open again.[63]

The highway engineers briefly considered placing a water and sewer line in the roadway, but the cost was prohibitive.[64] They also considered other routes, in order to avoid the Narrows. "The agonizing part of the Narrows is that a 100-year flood could cause great damage—not to property, but to the road," commented Governor Lamm. He continued, "It is not irrational to say that if we have to lose the road every 100 years, it is not a large risk to take. The problem is habitation and we have to make sure we don't expose those people to that risk."[65]

The Highway Department proposed five routes for U.S. 34. Various citizens' groups, such as the Big Thompson Action Group, Big Thompson Advisory Committee, and the Big Thompson Restoration Committee, strongly endorsed the old location, as did the editor of the Estes Park *Trail-Gazette* and the Estes Park Chamber of Commerce.[66] The old route was shown to be shortest, easiest, cheapest, and quickest to build. It would have the least harmful impact on wildlife and the economy. Best of all, the federal government would pay the total cost, estimated at $10,000,000. A different route, apparently, would obtain only ninety percent federal funding.[67] The Highway Department agreed and went to work.[68]

In order to construct private bridges, the residents organized a bridge district. For the twenty-three bridges repaired and built, the landowners paid less than a third of the cost. The bridges became county property. Thus, if damaged in the future, they would be restored with public money.[69] By Fall, 1978, the bridges were in place and the road repaired except for reinforcing the Narrows to withstand a 100-year flood. Thus far the cost was $8,500,000. The two-mile stretch in the Narrows would cost an additional $7,000,000.[70] The road was in the same place and looked much as it had before the flood. Considering the destruction caused by the storm, an

The Bureau of Reclamation replaced the siphon by the end of October. The supports had been knocked down by the flood.

observer might well question the wisdom of rebuilding in the same location.

Even the siphon was back as though nothing had occurred. There were some misgivings about the replacement. Clara Bell of Loveland wrote to the editor of the Loveland *Reporter-Herald,* "Only one good thing has come out of the terrible tragedy of the Big Thompson, and that is the washing away of that unsightly siphon at the entrance of the canyon." She hoped this "affront to the sensibility" did not return.[71] But it did return, and very quickly.

The Bureau of Reclamation paid Eagle Construction Company of Loveland $187,000 to replace the siphon. The material for the work, 220 feet of 108-inch steel pipe, came from Denver at a cost of $104,000. The conduit was in place by the end of October. The greatest noticeable difference was that it was painted green; the old one was silver.[72] At least the color was more compatible with the surroundings. Horsetooth Reservoir then began to refill, but the Bureau had to lower it to three percent of capacity the next year to repair an inlet. A mudslide had covered the inlet while the siphon was inoperative.[73] The repair work for the siphon was done so rapidly that there was little chance for anyone to suggest an alternative, as Clara Bell hoped. People, moreover, had other things to worry about.

Following a disaster, it is common for survivors to blame others for their losses and seek restitution. There were rumors, for example, that cloud seeding had caused the downpour. But no state permits had been issued, and the suspicion died quickly.[74] During the emergency phase of a catastrophe almost nothing is denied a victim in the way of food, shelter, clothing, and community support. Actually, the survivor is happy to have escaped death, and demands relatively little. Time passes, newspaper stories about the event drift into the back pages, and people return to the routine of living. There is less conversation about the disaster and, as the days go by, it fades into the history and folklore of the region. The victims, meanwhile, begin to assess their losses. They make demands and expect quick restitution. They had received it before, but now public interest has turned away. This provides a climate for acrimony and bitterness.[75] It also happened at the Big Thompson.

In mid-August some 250 to 300 survivors gathered at the Loveland Elks Lodge to hear officials speak about relief plans. It was an angry affair. People emotionally protested the 6-5/8 percent interest charged by the Small Business Administration, damned the limited emergency grants, and bristled at the suggestion of land expropriation. A leader, Joe Creek, stood and pointed to a short gray-haired woman who had lost her son and all her possessions. "When you go to get aid for us," he told the officials, "look at this lady and think what you're doing. Don't just give these people loans. Give them something they can live with, something they can be proud of."[76]

What should be given? In what amounts? How? The answers were unclear; there was little agreement. Unlike the evacuation mission which Englebert carried out efficiently in autocratic fashion, the questions of recovery required political settlement. Decisions were not dictated; they could not be. They were made through debate, hearings, persuasion, and votes. There was a search for what was morally and legally correct.

While the siphon was undergoing repair the level of Horsetooth Reservoir dropped.

The fundamental question involved private property in the canyon. Larry Timm, the Larimer County planner, said, "We're all of the opinion that we don't want that kind of a disaster to occur again. The only reason this was a disaster was that people lived there."[77] But what do you do with the people who are there, and what do you do about their property? The landowners wanted governmental support and unfettered use of their real estate in the canyon. But why should society pay the bill for such freedom? Another flash flood would come. Why not prevent the loss of life and property, if possible? That was the dilemma. There was a solution, but it took time to work it out.

Toward the end of the first week following the disaster, Governor Lamm again toured the canyon and met with officials. Donald Barnett, mayor of Rapid City in 1972, accompanied the governor. He had been through a similar situation. Barnett warned of the long recovery period and advised local politicians to work closely with the "muscle" of state and federal governments. The answer for Rapid City was to clear a floodplain park where no one was allowed to build. Rapid City bought the land with federal money. It was something the government had done before, but it was not a widespread practice.[78]

With gubernatorial consent, the Larimer County Commissioners declared a six-month "moratorium" on building in the floodplain. Within the high-water mark of the flood, no new structure could be built, and no structure over half destroyed could be restored. Repairs costing over $100 required clearance. Photographs from four sides or an official on-site inspection were necessary. Although an appeal to a review board was possible, the "moratorium," by and large, thwarted the tendency of people to rebuild in the same spot, in defiance of nature.[79] There was some slippage and a few persons obtained approval before the policy was set. One woman, for example, was allowed to build a flood wall without consideration that diverted water might damage a neighbor's land. The Federal Insurance Administration protested these violations and

The siphon under repair.

The new siphon is shown in place in 1978. U.S. 34 has also been restored.

Survivors of the disaster met with relief officials at one-stop centers in Loveland and Estes Park. *Photograph courtesy of Estes Park* Trail-Gazette.

they were stopped.[80]

The six-month restriction on building bought time. Meanwhile, the F.I.A. pursued a flood plain study of the Big Thompson, in order to provide maps. Without the insurance, which the F.I.A. could provide, land in the area was nearly worthless. The agency would not insure unless the county enacted floodplain zoning restrictions. Hence, they needed the maps. During the interim, residents gained some public representation, at first through the Big Thompson Advisory Committee. This ad hoc committee emerged after some of the angry participants at the Elks Lodge volunteered to pursue the residents' complaints. There was a question of legality and so the committee, in time, changed to the Big Thompson Recovery Planning Council which possessed local, state, and federal representation.[81]

Citizens also met sporadically with visiting officials and service groups to air complaints. The Interfaith Task Force turned up displeasure over S.B.A. guidelines, the high cost of bridges, and misinformation. Self-employed persons had been told they did not qualify for unemployment payments. This was untrue, but by the time they were properly informed, the deadline for application was past. Schelling said, "There just isn't a lot of consistency in the way decisions were made."[82] A somewhat humorous example came from an old man at Drake. He laughingly explained to the governor why he rejected an S.B.A. loan, "I didn't believe even the government would offer a 75-year old man $35,000 for thirty years." He wanted direct aid, not a loan.[83]

If massive federal aid was to be gained, application had to come from the county commissioners and the governor's office. Both appointed

Donald Barnett seated to the right of Governor Lamm gave advice to officials about the establishment of a floodplain park. He was mayor of Rapid City in 1972 when it was struck by a flash flood. This meeting took place at the end of the first week after the disaster and was significant for directing a solution to the problems of land use. *Photograph courtesy of Loveland* Reporter-Herald.

special staff people to the task. Lamm assigned Kenneth Wright and the commissioners hired Willard R. Quirk, formerly with a Greeley engineering firm. These two represented their governments at various meetings, particularly those of the Big Thompson Recovery Planning Council.

In late December, 1976, the F.I.A. presented the floodplain maps to the county commissioners who, in turn, invited a citizens' review. The F.I.A. guidelines prohibited construction within the 100-year floodway level. They defined "floodway" as eighteen inches of water moving three feet per second. The insurance rates, federal or private, within that area were so high that building could not possibly take place. Of interest, however, is the F.I.A. choice of a flood level. The Big Thompson disaster was one that would occur every 333 years, supposedly. The F.I.A. level was more than two-thirds less than what had actually occurred.[84] It was a compromise, and no one challenged the logic of the choice.

Residents searched through the maps and wondered whether or not they could rebuild. Seventy to eighty people could not. Most found their land half in and half out of the prohibited zone. Frank Hershman, for instance, discovered that his meadowland was inside and his mountain slope was outside the floodway. "The land I thought was worthless has become more valuable, but the land I thought was valuable is worth nothing," he said.[85] The commissioners held off a decision about zoning for a month and sent out 1,100 notices to property owners, to encourage inspection of the maps. There could be errors because they were based upon aerial photographs. Ground-level examination might reveal something different.[86]

In February, the three commissioners, Nona Thayer, David Weitzel, and William Lopez—Thayer and Weitzel were newly elected—approved a zoning ordinance for the Big Thompson area. It prohibited new construction in the 100-year floodway. Any building already there and less than fifty percent destroyed could remain; the rest could not. This made the property of eighty families valueless.[87] The residents opposed the ordinance and questioned its legality, but with an attitude of resignation. "We've been told we have to do these things right away or dire things will happen to us. Dire things already happened," said one. Ray Drake of Loveland who worked on debris removal argued on the contrary:

> I saw the grief and destruction first hand. . . . We know the consequences of no adoption. That's another tragedy. There will be another flood. . . . The question is when will it strike. The risks are now well understood and we (owe) the future owners of the property notice of the risk.

The logic behind the ordinance was apparent, but the loss of property was, as another resident said, "Pouring salt on the wound."[88]

The county had destroyed land value with its ordinance. Did it owe anything to the owners? Although admitting a moral obligation, George Hass, a county attorney, said, "In my opinion, no compensation has to be paid as a result of zoning."[89] Weitzel said to the owners, "If we don't take something away from you, (you) won't get paid for it. . . . If we don't approve this, (there is) no basis for claims of compensation." Thayer commented, "We have very little choice. It is a very 'iffy' proposition, unfortunately."[90]

The commissioners promised to work for compensation, and the goal was payment for land at pre-flood valuation. The county could not afford this rate of compensation, and the state predicted an $18,000,000 shortfall in tax revenue for 1977. The only hope was federal money, like that given to Rapid City. There were, unfortunately, cross-currents, disharmony, and a tendency to blame others rather than to cooperate.

During the move toward zoning, Governor Lamm met with forty residents at Drake. He asked if they thought people should be allowed to move back into the floodplain, and received, in response, a resounding, "Yes!" Lamm, however, was basically in favor of land-use planning and courageously retorted, "I think you're being selfish at the expense of future lives." He also chided the Larimer commissioners and said, "They can't get their act together."[91]

Reverend Schelling of Interfaith said of the residents, "They feel abandoned. What they really hurt from the most are statements by legislators and the public who blame them for the disaster because they lived in the canyon."[92] James Johnson, Colorado's congressman from the area, bristled about the situation, "I don't think the federal government is the villain in all this. Why should the federal government take any blame for saying we won't insure anyone who builds in a floodway? That's the major trouble with this."[93]

Nona Thayer found everything circular. As a commissioner, she could promise nothing because neither the state nor federal government would commit itself. Senator Gary Hart considered compensation to be a constitutional issue under the Fourteenth Amendment—a person cannot be deprived of property without due process of law—but he could make no promises. Governor Lamm cautioned, "The federal people can't insure against all natural disasters. . . . What happened in the Big Thompson Canyon was an act of God, an unkind one, but nevertheless not something you can point to anyone and say, 'You pay.'" The state, likewise, could make no promises to Big Thompson residents.[94]

Finally, the pieces of the puzzle came together. In December, 1976, the old commissioners sent a package of eight proposals to the state legislature for consideration. They were not specific enough, however, and there was no action. Legislator James Lloyd said, "I do not think any of the commissioners has had a handle (on) where they are going."[95] The commissioners did not hire Willard Quirk as a coordinator until October, and a county election intervened.

After the first of the year, the new commissioners presented a second package prepared by Quirk and the Big Thompson Recovery Planning Office. It requested $17,000,000 for land acquisition, a reduction to one percent for S.B.A. loans, forgiveness of federal mortgages, and $2,500,000 for relocation and home rehabilitation. Along with money for bridges and soil conservation, it amounted to $21,000,000 in federal aid.[96] In conjunction with this package, Fred Anderson, president of the Colorado Senate and a Loveland resident, planned to introduce a bill for $862,000 to help match the money for land acquisition. Ronald Strahle, a state representative from Fort Collins and a leader of the Colorado House, said pessimistically, "There is just no way you are going to get that kind of money from the legislature. People in other parts of the state don't understand the magnitude of the problem."[97]

Despite Strahle's pessimism, the aid began to come. National representatives James Johnson and Floyd Haskell obtained congressional approval to acquire land at pre-flood value. Federal law allowed purchase only at current prices, and land in the Big Thompson had been devalued by 90 percent. This acquisition, therefore, would be an exception. In the process of obtaining this legislation, Johnson again sniped, "They (state officials) evidently feel they don't have much of an obligation to the flood victims. The state does have an obligation. They should not foist these things off on the federal government. It's not totally a federal obligation."[98]

President Carter assented to payment at pre-flood prices and set up a program through Cecil Andrus, Secretary of Interior. Andrus awaited a new budget year and then authorized the flow of funds through the Bureau of Outdoor Recreation, which had a double-sized allocation for Colorado that year. "When the state officials get all their thinking together, we'll help them out," he said in a Colorado press statement. "The Big Thompson is typical of western canyons. I agree with your governor. It is a matter of not building in the floodplain."[99]

The pieces, thus, dropped into place for land acquisition: $1,012,600 from the Land and Water Conservation Fund of the Bureau of Land Management; $731,000 from the Colorado State general fund; $131,600 from Larimer County; and $150,000 from the Four Corners Regional Commission.[100]

There was a brief flare up between Governor Lamm and Fred Anderson over the Colorado portion. Anderson initially tacked on a rider to limit the authority of the state's Land Use Commission. The Commission had briefly enjoined the county zoning plan to prohibit overnight camping in the floodplain. L.U.C. officials estimated that thirty-four of the dead had been camping. Anderson, a Republican, struck at the Commission, which was close to the heart of the Democratic governor. Lamm vetoed the Big Thompson appropriation bill and suggested the passage of one without the rider. After a partisan debate, the bill passed without Anderson's provision, and the money became available.[101]

The funds bought 156 acres for a fifteen-mile county park through the floodway. The Big Thompson Recovery Planning Office negotiated the purchase, and closed its doors on June 30, 1978.[102] Although frustrated by the delay, a sample of landowners appeared satisfied. Some still wanted to go back, but all agreed they had received a fair price for their land.[103]

It was over. The Highway Department still had some on-going work for U.S. 34 in the Narrows, but for the people, the county, and the state of Colorado, the nightmare of the Big Thompson disaster was over.

Research among the survivors of the Big Thompson disaster resulted in this simple and direct advice for motorists in Colorado canyons.

Chapter V

Lessons: The Next Time

Two years after the Big Thompson disaster, Richard Lamm dedicated a memorial to Hugh Purdy and Michel Conley at the spot where road-workers found Purdy's crushed car. The Governor said, "Memorials tell us not to forget, and there are a few things we ought not to forget. The first is the sacrifice these officers made." He went on to assert that Colorado needed to be prepared. It was not an idle statement. Earlier that day, during a discussion period at the Foothills Unitarian Church in Fort Collins and later to the press, he commented that the state was only "marginally" better prepared for disaster than it had been before the Big Thompson flood.[1] That was a disturbing revelation. Was anything learned from the experience?

The Big Thompson catastrophe, by and large, confirmed findings about human behavior in similar situations: panic was infrequent, except in the face of danger; people acted positively and rationally to protect families and others; self-help was frequent, and the victims themselves carried out the initial rescue work; looting was rare, but there was a problem of convergence of volunteers; and disaster creates problems, but not chaos.[2] The Big Thompson also illustrated the importance of preparation and the long-term nature of recovery, with all of the political ramifications.

The study by Eve Gruntfest from the University of Colorado underscored the simple, but life-saving, action needed to escape a flash flood in the mountains — abandon cars and houses and climb the mountainside. The Highway Department erected large green and white signs with that advice along the route of the Big Thompson. The signs might make a difference — to someone — the next time.

The National Weather Service also reacted to improve its response. As a result of N.O.A.A. criticism, the N.W.S. moved to repair its local observer network on the Big Thompson, St. Vrain, and Poudre rivers. They recruited Bureau of Reclamation workers at Estes Park who already monitored rainfall for the Colorado-Big Thompson Project.[3] The N.W.S., in the future, should perform more quickly and with greater precision.

The number of flood insurance policies increased from 36 to 202 the year after the flood. This is both helpful and detrimental. It will help people if they are struck again, but since it is available, it tends to encourage people to live in the canyon. The highway was rebuilt in the same location with federal financing. That also tends to orient life in the same pattern as before. Larimer County, however, courageously zoned the floodway. Prevention is certainly the best action to take, but the plans are based on a supposed 100-year flood level. Nature cannot be so trusted, especially when there is an example of a flood over three times that size. The facts, rather than statistical estimates and compromises, should take precedence. The county should plan for the

Two years after the flood Governor Lamm dedicated a memorial to Hugh Purdy and Michel Conley at the site where Purdy's car was uncovered.

highest level of flood. At least, from other experience, that is the way to avoid tragedy.

There are examples. In 1936, the Connecticut River rose eight and a half feet higher than it had in 300 years. Two years later, its flood crest was six feet higher.[4] Thomas Ten Eyck who was head of the Department of Natural Resources in Colorado during the administration of Governor John Love told the following story from his younger years as an engineer:

> There had been some disastrous flooding on the Rio Grande and during the course of our design . . . the hydrology people in the Bureau had prepared what we called a spillway design flood. This is a kind of flood that shouldn't happen more frequently than once in a thousand years. It's a kind of a statistical guess if you don't have a lot of records.
>
> But during the course of the period when we were doing this design, we had a flood, an actual flood, that exceeded the thousand year design flood. Well, this sent the hydrologist back to the drafting board, so to speak. The following year we had another flood that exceeded the previously thought-to-be-thousand-year design flood. So we significantly enlarged the design flood and significantly enlarged the spillway.[5]

Rainstorms, in addition, occur in random fashion. As Victor Koelzer, a Colorado State University civil engineer, explained, "If you flip a coin nine times and each time it turns up heads, there is still a fifty-fifty chance the next time it will be tails." In other words, there is an equal chance for a Big Thompson flood every year. It could even be worse; other Colorado storms have produced greater rainfall.[6]

On the other hand, since the storms strike at random, the chances are slight that it will happen again for another 333 years. But it is a gamble, and the frightening fact is that Colorado is quite vulnerable in other places. A 1978 report by the Division of Disaster Emergency Services listed fourteen dangerous streams: Boulder, Clear Creek, Bear Creek, Monument, Big Thompson, St. Vrain, Gunnison, South Platte, Poudre, Buckhorn, Crystal, Cimarron, Rifle, and Roaring Fork. A flood on the South Platte similar to the one of 1965, the report pointed out,

One result of the flood has been to warn campers.

would do more damage today than it had before.[7]

For potential loss of life, another study revealed, Boulder Canyon is most exposed.[8] What if the stormcloud of the Big Thompson had formed and released its havoc fifty miles to the south? The leaders in Boulder are not unaware of the risk. Immediately following the Big Thompson flood, the Boulder Regional Communication Authority installed an electronic alarm in Boulder Creek three miles from the city limits. A twelve to fourteen inch rise would trigger the alarm. Of course, a wave of water which might occur in a flash flood would take it out. Even if it did work, the warning time would be insufficient. Flash floods move too quickly. To make matters worse, Boulder officials noted that the Big Thompson demonstrated an overflow three to four times greater than planned for in their own floodplain zoning.[9]

The Big Thompson Canyon is better prepared, of course, for the next time. The floodway is clear of homes and the Sheriff's Office has experience. The rest of Colorado is only slightly better off than before the Big Thompson storm. Not much was learned. There is a lack of rigorous land-use planning and a need for better development of systems for communication, body identification, and long-term recovery. Most of all, the state needs to reaffirm that human beings must live in harmony with the patterns of nature. Colorado, a state of intense beauty, carries that lesson on the face of the land. Human beings, in their ignorance and arrogance, should not ignore it. The mountains are still unforgiving. They require respect.

Appendix

Deaths in the Big Thompson Disaster

Name	Age	Residence
Adams, Gregory	27	Loveland, Colorado
Applebaum, Joseph	68	Lakewood, Colorado
Applebaum, Norma M. J.	68	Lakewood, Colorado
Bashford, Stannis	17	Loveland, Colorado
Back, Irene	48	Denver, Colorado
Back, William Ray	55	Commerce City, Colorado
Bailey, Keith	50	Big Thompson, Colorado
Becker, Louise	75	Waverly, Iowa
Bell, Eleanor	51	Drake, Colorado
Bell, Myrl	52	Drake, Colorado
Bissing, Terri	27	Glenwood Springs, Colorado
Bokelman, Frances	62	Drake, Colorado
Brooks, Dora	40	Austell, Georgia
Brooks, Elbert	41	Austell, Georgia
Campbell, Elaine	20	Dearborn, Michigan
Carlson, Stanley J.	22	Greeley, Colorado
Carlson, Tanya	19	Greeley, Colorado
Carr, James M.	21	Loveland, Colorado
Carter, Emery	64	Belleville, Illinois
Carter, Etta Marie	56	Belleville, Illinois
Channer, Genievieve H.	62	Drake, Colorado
Channer, Glenn	65	Drake, Colorado
Conley, Michel O.	38	Colorado Springs, Colorado
Curtis, Larry	25	Sterling, Colorado
Decker, Verna	91	Drake, Colorado
Eyl, Norman W.	58	Big Thompson, Colorado
Felts, Mary Jean	29	Springfield, Illinois
Fujiwara, Jane	27	Honolulu, Hawaii
Gass, Mildred	50	Denver, Colorado
Goddard, Richard	19	Denver, Colorado
Graham, Beverly	33	Loveland, Colorado
Graham, Clara	64	Cedar Cove, Colorado
Graham, Lisa	2	Loveland, Colorado
Greenlee, Vulah	84	Big Thompson, Colorado
Haskell, Chad	5	Des Moines, Iowa
Haskell, Emma	63	Des Moines, Iowa
Haskell, Judy	31	Des Moines, Iowa
Haskell, William	32	Des Moines, Iowa
Haskell, William F.	69	Des Moines, Iowa
Harmon, Greg	19	Keokuk, Iowa
Hazzard, Clarence	73	Austin, Texas
Henry, Frances	37	Drake, Colorado

Name	Age	Residence
Henry, George	51	Drake, Colorado
Henry, Laurie Lee	9	Drake, Colorado
Henry, Shirlee Ann	14	Drake, Colorado
Hobbs, Dorothy	66	Drake, Colorado
Hobbs, Ellis	66	Drake, Colorado
Horton, Kimberly	18	Aurora, Colorado
Huffsmith, Richard	19	Loveland, Colorado
Hufnagle, Ken	66	Loveland, Colorado
Hufnagle, Nora	67	Loveland, Colorado
Jenkins, Helen E.	61	Denver, Colorado
Jobst, Dorothy	52	Omaha, Nebraska
Jobst, Eugene (Rev.)	57	Omaha, Nebraska
Johnston, Rae Ann	28	Crystal, North Dakota
Karstens, Debra K.	22	Loveland, Colorado
Karstens, Donald D.	23	Loveland, Colorado
Koerner, Kurt	20	Arvada, Colorado
Kronenberger, Ed	51	Littleton, Colorado
Kronenberger, Esther	51	Littleton, Colorado
Kronenberger, Kim	22	Littleton, Colorado
Lafferty, Mildred	58	Drake, Colorado
Lafferty, Pat	60	Drake, Colorado
Lee, Ealy A.	62	Loveland, Colorado
Lemon, Kelly	13	Des Moines, Iowa
Leyden, Barbara	29	McLean, Virginia
Lohry, Andrew	72	Drake, Colorado
Lohry, Elizabeth	47	Drake, Colorado
Loomis, Cathie M.	29	Seattle, Washington
Lorence, Brenda	15	Phillipsburg, Kansas
Lorence, Debra	13	Phillipsburg, Kansas
Lorence, James Lyle	43	Phillipsburg, Kansas
Lorence, Phyllis	41	Phillipsburg, Kansas
Maciejewski, Benjamin L.	48	Big Thompson, Colorado
Maciejewski, Betty	50	Big Thompson, Colorado
Manongdo, Priscilla	28	Quezon City, Philippines
Mares, Alyson Dell	5	Big Thompson, Colorado
Mares, James, Jr.	33	Big Thompson, Colorado
Mares, Jamie	8	Big Thompson, Colorado
Mares, Renee (Mrs.)	33	Big Thompson, Colorado
Mares, Ricky	6	Big Thompson, Colorado
Marshall, Eulah	77	Wichita, Kansas
Matthews, George	17	Aurora, Colorado
Matthews, Katherine	54	Aurora, Colorado
McCargo, Lois	63	Big Thompson, Colorado
McKissack, Jack	58	Aurora, Colorado
Miller, Jeffrey	27	Greeley, Colorado
Morgan, Homer M.	21	Aurora, Colorado
Morningstar, Dorothy	64	Lisbon, Iowa
Morningstar, Leon	65	Lisbon, Iowa
Moore, Debora	17	Loveland/Drake, Colorado
Morton, Charles E.	62	Estes Park, Colorado
Nelson, Myrtle	45	Fleming, Colorado
Nelson, William I.	48	Fleming, Colorado
Ogley, Tena Marie	19	Ovid, Colorado
Oler, Nettie	52	Longmont, Colorado

Name	Age	Residence
Palmer, Nona Jean	19	Fitzsimmons-Denver, Colorado
Plant, John Joseph	42	Broomfield, Colorado
Plant, Judy	37	Broomfield, Colorado
Plant, Michael	14	Broomfield, Colorado
Primm, Grace	79	Kansas City, Missouri
Purdy, W. Hugh	53	Loveland, Colorado
Remsing, Francis	48	Big Thompson, Colorado
Remsing, Martin A.	8	Big Thompson, Colorado
Remsing, Martin J.	65	Big Thompson, Colorado
Rhoad, Carol	23	Grantsville, Pennsylvania
Richardson, Elizabeth	54	Springfield, Missouri
Richardson, Leonard	59	Springfield, Missouri
Richardson, Lisa	16	Springfield, Missouri
Richardson, Rebecca	9	Springfield, Missouri
Rogers, Ellen	58	Brighton, Colorado
Rogers, Wes	59	Brighton, Colorado
Rudd, Cheryl	19	Greeley, Colorado
Saunders, Eugene	65	Lawrence, Kansas
Saunders, La Vera Faith	61	Lawrence, Kansas
Scavone, Irene	77	Oak Lawn, Illinois
Schell, Claude	85	Greeley, Colorado
Schell, Mae Loretta	70	Greeley, Colorado
Schlater, Catherine	23	Denver, Colorado
Seibert, Annie	61	Caldwell, Texas
Seibert, Casper L.	63	Caldwell, Texas
Sheets, Rex	26	Loveland, Colorado
Smith, Earl R.	30	Drake, Colorado
Smith, Jacob	3	Elgin, Illinois
Smith, Joseph S.	64	Drake, Colorado
Smith, Lylith	57	Lexington, Tennessee
Smith, Sara	5	Elgin, Illinois
Starr, Ann	79	Loveland, Colorado
Stinnett, Shirley	56	Kansas City, Missouri
Turner, Drew	25	Fort Collins, Colorado
Venrick, Mae	50	Catawba, Ohio
Venrick, Ross	61	Catawba, Ohio
Waters, Clarence B.	64	Riviera, Arizona
Waters, Mary Sue	62	Riviera, Arizona
Watson, Cheryl	21	Loveland, Colorado
Watts, Aaron	3	Fort Collins, Colorado
Weiss, Beulah	65	National City, California
Woodring, Edna May	57	Estes Park, Colorado
Woodring, Fred	63	Estes Park, Colorado

Endnotes

Chapter I

1. Ansel Watrous, *History of Larimer County* (Fort Collins, Colorado: Courier Printing, 1911), pp. 170-171. Erme Burton Stoner, "History of Larimer County," *Historical Encyclopedia of Colorado* (Denver: Colorado Historical Association, 1975), p. 259. *Triangle Review* (Fort Collins), August 2, 1978.
2. F. V. Hayden, *Preliminary Field Report of the United States Geological Survey of Colorado and New Mexico* (Washington, D.C.: Government Printing Office, 1869), pp. 20-26. Robert Ormes, *Guide to the Colorado Mountains* (sixth edition; Chicago: Swallow Press, 1970), pp. 2-3.
3. Ormes, *Colorado Mountains,* pp. 21-22.
4. Fort Collins *Coloradoan,* special edition, August, 1976.
5. Toups Corporation, "Big Thompson Disaster Recovery Planning Report" (Loveland, Colorado; May 10, 1977), p. 61.
6. Marguerite Lindsey, compiler, *History of the Big Thompson Canyon* (Drake Home Demonstration Club, 1940), p. 2. Zethyl Gates, "Once Known as Loveland Grand Canyon," *The Big Thompson Disaster,* Grant Judkins, editor (Loveland: Lithographic Press, 1976), no paging. Fort Collins *Coloradoan,* special edition, August, 1976. *Triangle Review,* August 11, 1976.
7. Lindsey, *Big Thompson Canyon,* p. 2.
8. *Ibid.,* p. 5.
9. *Triangle Review,* August 23, 1978.
10. Loveland *Reporter-Herald,* March 26-27, 1977.
11. Fort Collins *Coloradoan,* special edition, August, 1976.
12. Estes Park *Trail-Gazette,* August 18, 1976.
13. Harry Hansen, editor, *Colorado: A Guide to the Highest State* (new revised edition; New York: Hastings House, 1970), p. 214.
14. Toups Corporation, "Planning Report," pp. 56-59.
15. Corps of Engineers, Department of the Army, *Flood Plain Information: Big Thompson River, Little Thompson River* (June, 1977), p. 2.
16. Corky Rogers, Oral History Interview, September 8, 1976, pp. 27-29.
17. Corps of Engineers, Department of the Army, *Flood Plain Information: Big Thompson River, Loveland, Colorado* (December, 1971), pp. 10-11.
18. Gingery Associates, Inc., *Special Flood Plain Information Report: Big Thompson River and Tributaries, Larimer County, Colorado* (Englewood, Colorado: December, 1976), p. 6.
19. Gates, "Loveland Grand Canyon," in Judkins, *Big Thompson Disaster.*
20. Loveland *Reporter,* June 7, 1894.
21. Gates, "Loveland Grand Canyon," in Judkins, *Big Thompson Disaster.* Loveland *Daily Herald,* August 1, 2, 1919.
22. Loveland *Daily Herald,* June 1, 1921.
23. *Ibid.,* June 6, 1921.
24. *Ibid.,* June 9, 1921.
25. George P. Cressman, "Flash Flood Warnings—Federal Plus Local Action," Natural Hazards Observer, 3 (September, 1978), 1. Denver *Post,* June 27, 1978.

26. Carl Ubbelohde, Maxine Benson, Duane A. Smith, *A Colorado History* (third edition; Boulder: Pruett Press, 1972), p. 85.
27. Denver *Post,* August 2, 1976.
28. Don Hataway, Oral History Interview, September 10, 1976, pp. 34-35.
29. Logan Rappe, Oral History Interview, September 23, 1976, pp. 2-7.
30. Denver *Post,* August 4, 1976.
31. Corps of Engineers, *Flood Plain Information* (1971), pp. 10, 16.
32. Denver *Post,* August 2, 1976.
33. *Ibid.*, August 16, 1976.
34. *Ibid.*
35. *Ibid.* Gingery Associates, *Flood Plain Information Report,* p. 18.
36. John Englebert, Oral History Interview, August 18, 1976, pp. 1-3. Rich Hovey, Oral History Interview, August 2, 1976, pp. 2-4.
37. Robert Watson, Oral History Interview, August 17, 1976, p. 2.
38. David Essex, Oral History Interview, September 10, 1976, pp. 1-2.
39. "The Real Tragedy Is People," Grant Judkins, editor, *The Big Thompson Disaster* (Loveland: Lithographic Press, 1976), no paging.
40. Loveland *Reporter-Herald,* August 20, 1976.
41. James L. Ross, "The Salvation Army," *American Behavioral Scientist,* 13 (January-February, 1970), 406, 412.
42. William D. Weller, Oral History Interview, December 13, 1976, pp. 2, 12, 18, 29-30.
43. Arnold R. Parr, "Organizational Response to Community Crisis and Group Emergence," *American Behavioral Scientist,* 13 (January-February, 1970), 423-429.

Chapter II

1. Robert A. Maddox, *et al.*, *Meteorological Aspects of the Big Thompson Flash Flood of 31 July 1976* (NOAA Technical Report ERL 388-APC L 41, July, 1977; Washington, D.C.: U.S. Government Printing Office, 1977), p. 3.
2. *Ibid.*, p. 1. Fort Collins *Coloradoan,* special edition, August, 1976.
3. Maddox, *Meteorological Aspects,* p. 63. Fort Collins *Coloradoan,* special edition, August, 1976. Robert C. Bolin, "Family Recovery from Natural Disaster: The Case of Rapid City" (unpublished Ph.D. thesis, University of Colorado, Boulder, 1976), pp. 32-33.
4. William Thomas, Oral History Interview, September 14, 1976, p. 6.
5. Fort Collins *Coloradoan,* special edition, August, 1976.
6. "The Real Tragedy Is People," Grant Judkins, editor, *The Big Thompson Disaster* (Loveland: Lithographic Press, 1976), no paging.
7. Toups Corporation, "Big Thompson Disaster Recovery Planning Report" (Loveland, May 10, 1977), p. 17.
8. Judkins, *Big Thompson Disaster.*
9. Thomas, Oral History Interview, pp. 6-8.
10. Toups, "Big Thompson Report," pp. 79, 82, 83, 90. Denver *Post,* September 29, 1976. Fort Collins *Coloradoan,* September 5, 1976.
11. Fort Collins *Coloradoan,* July 7, 1977.
12. Denver *Post,* November 13, 1976.
13. *Ibid.* Fort Collins *Coloradoan,* August 10, 1976.
14. Fort Collins *Coloradoan,* August 10, 1976. Denver *Post,* November 13, 1976.
15. Loveland *Reporter-Herald,* August 11, 1976.
16. *Triangle Review* (Fort Collins), August 11, 1976.
17. *Ibid.*
18. *KCOL Special Report* (no place, no publisher, no date), no paging.
19. *Ibid.*

20. Dennis S. Mileti, Thomas E. Drabek, J. Eugene Haas, *Human Systems in Extreme Environments: A Sociological Perspective* (Boulder: Institute of Behavioral Science, University of Colorado, 1975), p. 41.
21. Fort Collins *Coloradoan,* special edition, August, 1976.
22. *Triangle Review,* August 11, 1976.
23. Fort Collins *Coloradoan,* special edition, August, 1976. Denver *Post,* November 13, 1976. Tim Littlejohn, Oral History Interview, October 6, 1976, pp. 1-3.
24. Littlejohn, Oral History Interview, pp. 4-9.
25. *Triangle Review,* August 11, 1976. Judkins, *Big Thompson Disaster.*
26. Fort Collins *Coloradoan,* special edition, August, 1976.
27. Terry Urista, Oral History Interview, August 18, 1976, pp. 1-3.
28. *Ibid.,* pp. 3-4.
29. *Ibid.,* pp. 5-7.
30. Loveland *Reporter-Herald,* August 10, 1976.
31. *Ibid.,* August 9, 1976. Don Hataway, Oral History Interview, September 10, 1976, p. 27.
32. Judith Ann Miller, "Families in the Aftermath of Disaster: The Big Thompson Flood of 1976" (unpublished master's thesis, Colorado State University, Fort Collins, 1977), p. 24. Edith Hill Kimball, "Recovery of the Older Survivors of the 1976 Big Thompson Flood" (unpublished master's thesis, Colorado State University, Fort Collins, 1978), p. 33.
33. Eve C. Gruntfest, *What People Did During the Big Thompson Flood* (Boulder: Institute of Behavioral Science, University of Colorado, 1977), pp. 14-15, 22.
34. Rick Hays, Oral History Interview, August 24, 1976, pp. 23-25.
35. Urista, Oral History Interview, pp. 5-6. Gruntfest, *What People Did,* p. 6.
36. Judkins, *Big Thompson Disaster.*
37. Denver *Post,* August 5, 1976.
38. Mileti, Drabek, and Haas, *Human Systems in Extreme Environments,* pp. 42-44,
39. Denver *Post,* August 5, 1976.
40. Judkins, *Big Thompson Disaster.* Fort Collins *Coloradoan,* August 2, 1976.
41. Mileti, Drabek, Haas, *Human Systems in Extreme Environments,* pp. 57-58.
42. Ervin Deal, Oral History Interview, August 13, 1976, pp. 1-11. Fort Collins *Coloradoan,* August 2, 1976.
43. Denver *Post,* August 2, 1976.
44. Judkins, *Big Thompson Disaster.* Loveland *Reporter-Herald,* August 4, 1976.
45. Gary Haxton, Oral History Interview, August 24, 1976, pp. 1-4.
46. *Ibid.,* p. 6.
47. Fort Collins *Coloradoan,* August 2, 1976.
48. Denver *Post,* August 5, 1976.
49. Estes Park *Trail-Gazette,* August 13, 1976.
50. *Ibid.,* see letter to editor from Ruth Herndon.
51. Dorothy Ferguson, Oral History Interview, August 21, 1976, pp. 1-4.
52. Bill Wells, Oral History Interview, August 21, 1976, pp. 1-6. *Triangle Review,* July 13, 1977. Judkins, *Big Thompson Disaster.*
53. Loveland *Reporter-Herald,* August 4, 1976. Judkins, *Big Thompson Disaster.*
54. Fort Collins *Coloradoan,* August 6, 1976.
55. Loveland *Reporter-Herald,* August 10, 1976.
56. *Ibid.* Littlejohn, Oral History Interview, pp. 7, 12, 15-18.
57. *National Enquirer,* September 21, 1976.
58. Virginia McGee, Oral History Interview, August 22, 1976, pp. 1-5, 13.
59. Loveland *Reporter-Herald,* August 17, 1976. Judkins, *Big Thompson Disaster.*
60. Judkins, *Big Thompson Disaster.*
61. Richard Huffsmith, Oral History Interview, August 23, 1976, pp. 2-5, 12.
62. George, Rachel, and Greg Morgan, Oral History Interview, August 23, 1976, pp.-1-2, 4-7, 8-10, 18.
63. Urista, Oral History Interview, pp. 5-10, 15, 21.

64. *National Enquirer,* September 21, 1976. John McMaster and George Woodson, Oral History Interview, September 6, 1976, pp. 1-6.
65. McMaster and Woodson, Oral History Interview, p. 6.
66. *Ibid.,* pp. 7-36, 39.
67. Fort Collins *Coloradoan,* August 3, 1976.
68. Denver *Post,* August 2, 1976.
69. Loveland *Reporter-Herald,* August 12, 1976. Denver *Post,* August 2, 1976.
70. Denver *Post,* August 2, 1976.
71. *Ibid.*
72. Loveland *Reporter-Herald,* August 5, 1976.
73. Judkins, *Big Thompson Disaster.*
74. Mileti, Drabek, and Haas, *Human Systems in Extreme Environments,* p. 59.
75. Gruntfest, *What People Did,* pp. iii, 22, 23.
76. *National Enquirer,* September 21, 1976.

Chapter III

1. John Englebert, Oral History Interview, August 18, 1976, pp. 4-5.
2. Rich Hovey, Oral History Interview, August 12, 1976, pp. 6-8.
3. Englebert, Oral History, p. 9.
4. William Thomas, Oral History Interview, September 14, 1976, pp. 1-3.
5. *Ibid.,* pp. 4-5, 6, 9.
6. Fort Collins *Coloradoan,* special edition, August, 1976; August 2, 1976. Loveland *Reporter-Herald,* August 5, 1976.
7. Thomas, Oral History, pp. 7, 10. Englebert, Oral History, p. 7.
8. David Essex, Ronald Cotten, and Dwight Hamilton, Oral History Interview, September 10, 1976, pp. 2-9.
9. *Ibid.,* pp. 2-9. Grant Judkins, editor, *The Big Thompson Disaster* (Loveland: Lithographic Press, 1976), no paging. Lou Murphy, "A Night of Terror," *CB Magazine,* 96 (January, 1977), 27-29.
10. Judkins, *Big Thompson Disaster.*
11. Alvin Kruchten, Oral History Interview, August 20, 1976, p. 8.
12. Essex, Cotten, and Hamilton, Oral History, pp. 20-23.
13. *Ibid.,* p. 32. Estes Park *Trail-Gazette,* August 11, 1976.
14. Hovey, Oral History, pp. 10-15.
15. Thomas, Oral History, pp. 12, 17-18.
16. Gene Greenwalt, Oral History Interview, August 16, 1976, pp. 1-2.
17. Englebert, Oral History, pp. 9-10. Thomas, Oral History, pp. 12-14.
18. James Lloyd, Oral History Interview, October 8, 1976, pp. 4-8.
19. Richard D. Lamm, Oral History Interview, November 19, 1976, pp. 1-5. William D. Weller, Oral History Interview, December 13, 1976, pp. 1, 17. Fort Collins *Coloradoan,* special edition, August, 1976.
20. Thomas, Oral History, p. 18.
21. Arnold J. Parr, "A Brief View on the Adequacy and Inadequacy of Disaster Plans and Preparation in Ten Community Crises" (Disaster Research Center Paper, Ohio State University, Columbus, 1969), pp. 5-8.
22. Englebert, Oral History, pp. 11-12.
23. Lamm, Oral History, pp. 5-8.
24. Kruchten, Oral History, p. 7.
25. Fort Collins *Coloradoan,* special edition, August, 1976.
26. Logan Rappe, Oral History Interview, September 23, 1976. p. 14. Weller, Oral History, pp. 6, 18. Fort Collins *Coloradoan,* August 12, 1976.
27. Fort Collins *Coloradoan,* August 3, 1976.
28. Thomas, Oral History, pp. 58-60. Denver *Post,* August 2, 1976.

29. Greenwalt, Oral History, pp. 5-8, 11. Fort Collins *Coloradoan,* August 3, 1976. Denver *Post,* August 3, 1976.

30. George, Rachel, and Greg Morgan, Oral History Interview, August 23, 1976, pp. 12, 14. Englebert, Oral History, pp. 30-31.

31. Englebert, Oral History, pp. 12-13. Thomas, Oral History, p. 7. *Triangle Review* (Fort Collins), August 4, 1976. Loveland *Reporter-Herald,* August 5, 1976.

32. Loveland *Reporter-Herald,* August 3, 1976. Fort Collins *Coloradoan,* August 3, 1976.

33. Loveland *Reporter-Herald,* August 5, 1976.

34. *Ibid.,* August 9, 21-22, 1976.

35. Barbara O'Reilly, Oral History Interview, August 17, 1976, pp. 6-15, 18, 26-27, 30, 37. Thomas World, Oral History Interview, August 19, 1976, pp. 6, 13, 20-23, 26. Fort Collins *Coloradoan,* August 3, 9, 1976. Loveland *Reporter-Herald,* August 3, 1976.

36. Dennis S. Mileti, Thomas E. Drabek, and J. Eugene Haas, *Human Systems in Extreme Environments: A Sociological Perspective* (Boulder: Institute of Behavioral Science, University of Colorado, 1975), pp. 114-115.

37. Fort Collins *Coloradoan,* November 25, 1976.

38. James L. Ross, "The Salvation Army," *American Behavioral Scientist,* 13 (January-February, 1970), 406, 412. O'Reilly, Oral History, pp. 13a-13b.

39. Bill Birks, Oral History Interview, August 28, 1976, pp. 2, 4, 6-10. Mervyn L. Morelock, "The Big Thompson Flood," *The War Cry of the Salvation Army,* 96 (October 2, 1976), 20-21.

40. Englebert, Oral History, p. 24. Denver *Post,* August 3, 1976. Loveland *Reporter-Herald,* August 3, 1976. Fort Collins *Daily Journal,* August 5, 1976.

41. "The Real Tragedy Is People," Grant Judkins, editor, *The Big Thompson Disaster* (Loveland: Lithographic Press, 1976), no paging.

42. Englebert, Oral History, pp. 34-35. Denver *Post,* August 6, 1976.

43. Loveland *Reporter-Herald,* August 4, 5, 1976. Fort Collins *Coloradoan,* special edition, August, 1976. Judkins, *Big Thompson Disaster.*

44. Estes Park *Trail-Gazette,* August 6, 1976. Loveland *Reporter-Herald,* August 13, 1976.

45. Thomas, Oral History, p. 61. Rappe, Oral History, p. 19. Judkins, *Big Thompson Disaster.* Fort Collins *Coloradoan,* August 2, 1976.

46. Englebert, Oral History, pp. 17-18. Loveland *Reporter-Herald,* August 5, 1976. Denver *Post,* August 8, 1976.

47. Lloyd, Oral History, pp. 30-31. Kruchten, Oral History, p. 11. "The Real Tragedy Is People," Judkins, *Big Thompson Disaster.* Loveland *Reporter-Herald,* August 9, 1976.

48. Russell R. Dynes and E. L. Quarantelli, "Patterns of Looting and Property Norms: Conflict and Consensus in Community Emergencies" (Disaster Research Center Paper, Ohio State University, Columbus, 1968), pp. 10-15. Mileti, Drabek, and Haas, *Human Systems in Extreme Environments,* p. 119.

49. Richard Huffsmith, Oral History Interview, August 23, 1976, p. 14. Loveland *Reporter-Herald,* August 6, 1976.

50. Englebert, Oral History, pp. 17-18. Loveland *Reporter-Herald,* August 5, 1976.

51. Denver *Post,* August 2, 8, 1976. Fort Collins *Coloradoan,* August 4, 8, 12, 1976.

52. Fort Collins *Coloradoan,* August 2, 1976. Thomas, Oral History, p. 22. Mileti, Drabek, and Haas, *Human Systems in Extreme Environments,* pp. 87-88.

53. Englebert, Oral History, p. 16.

54. Denver *Post,* August 2, 1976.

55. Mentioned at a review of newspaper reporters at Colorado State University, October 27, 1976. Lloyd, Oral History, p. 51.

56. Englebert, Oral History, pp. 11, 23, 32.

57. *Ibid.,* p. 19. Fort Collins *Coloradoan,* September 17, 1976.

58. Denver *Post,* August 5, 1976.

59. Fort Collins *Coloradoan,* special edition, August, 1976. Denver *Post,* September 6, 21, 26, 1976.
60. Loveland *Reporter-Herald,* August 5, 9, 1976.
61. *Ibid.,* August 12, 14-15, 1976.
62. *Ibid.,* August 7-8, 1976. Greenwalt, Oral History, p. 6.
63. Fort Collins *Coloradoan,* August 13, 1976. Loveland *Reporter-Herald,* August 17, 1976.
64. Fort Collins *Coloradoan,* August 6, 1976.
65. *Ibid.*
66. Michael Charney, Oral History Interview, September 5, 1976, pp. 1-4, 10-13, 17-22, 24-28, 36-40. Fort Collins *Coloradoan,* special edition, August, 1976.
67. Charney, Oral History, p. 37.
68. Judith Ann Miller, "Families in the Aftermath of Disaster: The Big Thompson Flood of 1976" (unpublished master's thesis, Colorado State University, Fort Collins, 1977), p. 49. Edith Hill Kimball, "Recovery of the Older Survivors of the 1976 Big Thompson Flood" (unpublished master's thesis, Colorado State University, Fort Collins, 1978), p. 48.
69. *Rocky Mountain News,* September 3, 1976. Denver *Post,* August 3, 1976, September 2, 12, 1976.
70. Denver *Post,* August 5, 1976.
71. James Dooney, Oral History Interview, August 18, 1976, p. 19.
72. *Ibid.,* pp. 19-20. Judkins, *Big Thompson Disaster.* Denver *Post,* September 9, 1976.
73. Denver *Post,* November 7, 1976.
74. Dooney, Oral History, p. 14. Charney, Oral History, p. 18.
75. Fort Collins *Coloradoan,* August 3, 1976.
76. Denver *Post,* August 8, 1976.
77. Loveland *Reporter-Herald,* August 6, 1976.
78. Dooney, Oral History, pp. 11-12. Charney, Oral History, p. 19.
79. Dooney, Oral History, p. 22. Fort Collins *Coloradoan,* September 15, 1976.

Chapter IV

1. David Essex, Oral History Interview, September 10, 1976, p. 34.
2. Fort Collins *Coloradoan,* October 23, 1977; April 16, 1978. Loveland *Reporter-Herald,* August 7-8, 1976.
3. Loveland *Reporter-Herald,* July 2, 1976.
4. Gary Hart, Oral History Interview, October 28, 1976, pp. 4-9, 12-13.
5. Estes Park *Trail-Gazette,* August 20, 1976.
6. Fort Collins *Coloradoan,* special edition, August, 1976. Denver *Post,* August 2, 1976.
7. Fort Collins *Coloradoan,* special edition, August, 1976.
8. James Lloyd, Oral History Interview, October 8, 1976, pp. 36-40. Estes Park *Trail-Gazette,* August 25, 1976.
9. Fort Collins *Coloradoan,* August 4, 16, 1976. Estes Park *Trail-Gazette,* August 6, 1976.
10. Fort Collins *Coloradoan,* August 12, 1976; September 5, 1976; June 24, 1977. *Triangle Review* (Fort Collins), March 18, 1978.
11. Fort Collins *Coloradoan,* September 5, 1976. Denver *Post,* August 6, 1976. Loveland *Reporter-Herald,* August 7-8, 1976.
12. Fort Collins *Coloradoan,* September 5, 1976. Loveland *Reporter-Herald,* August 10, 1976.
13. Fort Collins *Coloradoan,* September 5, 1976.
14. Loveland *Reporter-Herald,* August 5, 1976. Denver *Post,* August 6, 1976.

15. Fort Collins *Coloradoan,* August 9, 1976. Loveland *Reporter-Herald,* August 13, 17, 1976.
16. Fort Collins *Coloradoan,* August 5, 1976. Loveland *Reporter-Herald,* August 11, 14-15, 1976. Estes Park *Trail-Gazette,* August 18, 25, 1976.
17. Loveland *Reporter-Herald,* August 18, 1976.
18. Fort Collins *Coloradoan,* September 6, 1978.
19. Denver *Post,* September 29, 1976.
20. *Ibid.*, September 1, 1976. Fort Collins *Coloradoan,* August 16, 1976. Loveland *Reporter-Herald,* August 17, 1976.
21. Loveland *Reporter-Herald,* August 18, 1976.
22. Rick Hays, Oral History Interview, August 24, 1976, pp. 5-8, 11, 16.
23. *Ibid.*, p. 22.
24. Fort Collins *Coloradoan,* March 4, 1977. *Triangle Review,* February 2, 1977. Loveland *Reporter-Herald,* August 4, 9, 1976.
25. Loveland *Reporter-Herald,* August 5, 10, 12, 14-15, 1976.
26. Denver *Post,* August 8, 1976; September 1, 1976.
27. Loveland *Reporter-Herald,* August 14-15, 1976. Estes Park *Trail-Gazette,* August 20, 1976. Sam Minter, Oral History Interview, August 23, 1976, pp. 10-12, 18.
28. Denver *Post,* September 7, 1976.
29. Estes Park *Trail-Gazette,* August 27, 1976.
30. W. L. Wells, Oral History Interview, August 21, 1976, pp. 14-18, 20, 22-25, 26-27.
31. *Triangle Review,* July 31, 1977.
32. Robert Schelling, Oral History Interview, September 14, 1976, pp. 1-5, 11, 19, 22.
33. Edith Hill Kimball, "Recovery of the Older Survivors of the Big Thompson Flood" (unpublished master's thesis, Colorado State University, Fort Collins, 1978), p. 61.
34. Fort Collins *Coloradoan,* December 22, 1976.
35. Loveland *Reporter-Herald,* August 10, 1976.
36. Estes Park *Trail-Gazette,* August 13, 1976. Loveland *Reporter-Herald,* August 14-15, 1976.
37. Loveland *Reporter-Herald,* August 14-15, 1976.
38. *Ibid.*, August 11, 1976.
39. Corky Rogers, Oral History Interview, September 8, 1976, pp. 2-3, 4, 7-8, 13-14.
40. Loveland *Reporter-Herald,* August 5, 1976.
41. Rogers, Oral History, pp. 12, 14, 15-18. Estes Park *Trail-Gazette,* August 13, 1976.
42. Denver *Post,* August 12, 1976.
43. *Ibid. Triangle Review,* August 11, 1976.
44. Denver *Post,* August 10, 1976.
45. Fort Collins *Coloradoan,* August 9, 1976.
46. Denver *Post,* August 10, 1976.
47. *Triangle Review,* March 23, 1977. Colorado Department of Revenue, Sales Tax Statistics for Counties and Selected Cities, 1978.
48. *Triangle Review,* August 11, 1976.
49. Denver *Post,* August 13, 1976. Estes Park *Trail-Gazette,* August 27, 1976.
50. Estes Park *Trail-Gazette,* August 20, 1976.
51. *Ibid.*, August 27, 1976.
52. *Ibid.*
53. Loveland *Reporter-Herald,* August 11, 1976.
54. *Ibid.*, August 7-8, 17, 1976.
55. *Ibid.*, August 13, 1976.
56. *Ibid.*, August 4, 1976. Don Hataway, Oral History Interview, September 10, 1976, pp. 8-12.
57. Hataway, Oral History, pp. 9-13. Loveland *Reporter-Herald,* August 17, 1976.
58. Fort Collins *Coloradoan,* August 16, 1976. Loveland *Reporter-Herald,* August 7, 1951.

59. Hataway, Oral History, pp. 18-20, 22. Loveland *Reporter-Herald,* August 11, 1976. *Triangle Review,* August 23, 1978.
60. Colorado Department of Revenue, Sales Tax Statistics for Counties and Selected Cities, 1978.
61. Loveland *Reporter-Herald,* August 3, 1976.
62. *Ibid.,* August 7-8, 1976.
63. Denver *Post,* October 26, 1976. Fort Collins *Coloradoan,* October 26, 1976.
64. Loveland *Reporter-Herald,* August 17, 1976.
65. Estes Park *Trail-Gazette,* August 13, 1976.
66. Loveland *Reporter-Herald,* August 17, 1976. Estes Park *Trail-Gazette,* August 18, 27, 1976. Fort Collins *Coloradoan,* August 23, 24, 1976.
67. Fort Collins *Coloradoan,* August 18, 1976; September 1, 1976. Denver *Post,* September 1, 1976.
68. Fort Collins *Coloradoan,* October 8, 1976.
69. *Ibid.,* January 25, 1977; January 17, 1978.
70. *Triangle Review,* August 2, 1978.
71. Loveland *Reporter-Herald,* August 18, 1976.
72. *Ibid.,* August 17, 1976.
73. Fort Collins *Coloradoan,* November 13, 1977. *Triangle Review,* January 12, 1977.
74. Denver *Post,* August 4, 1976. Loveland *Reporter-Herald,* August 10, 1976.
75. Dennis S. Mileti, Thomas E. Drabek, and J. Eugene Haas, *Human Systems in Extreme Environments: A Sociological Perspective* (Boulder: Institute of Behavioral Science, University of Colorado, 1975), pp. 100, 105, 107.
76. Denver *Post,* August 15, 1976.
77. *Triangle Review,* August 11, 1976.
78. *Ibid.* Fort Collins *Coloradoan,* August 6, 1976. Loveland *Reporter-Herald,* August 6, 1976. Denver *Post,* August 7, 1976. Mileti, Drabek, and Haas, *Human Systems in Extreme Environments,* p. 29.
79. Fort Collins *Coloradoan,* August 10, 1976. Estes Park *Trail-Gazette,* August 11, 1976.
80. Fort Collins *Coloradoan,* November 28, 1976; January 4, 1977.
81. *Ibid.,* August 15, 23, 1976; September 28, 1976; October 12, 1976; November 16, 1976. Loveland *Reporter-Herald,* August 16, 1976.
82. Fort Collins *Coloradoan,* March 4, 1977.
83. *Ibid.,* December 21, 1976.
84. *Ibid.,* December 22, 1976. Denver *Post,* August 16, 1976.
85. Fort Collins *Coloradoan,* December 23, 1976.
86. *Ibid.,* January 5, 1977.
87. *Ibid.,* January 27, 1977; February 1, 2, 1977.
88. *Ibid.,* February 1, 1977.
89. *Ibid.,* January 27, 1977.
90. *Ibid.,* February 1, 2, 1977.
91. *Ibid.,* December 21, 1976.
92. *Triangle Review,* January 12, 1977.
93. *Ibid.,* February 2, 1977.
94. *Ibid.,* January 26, 1977.
95. Denver *Post,* December 26, 1976.
96. *Ibid.,* January 15, 1977. Fort Collins *Coloradoan,* January 13, 1977.
97. *Triangle Review,* March 9, 1977.
98. *Ibid.,* June 1, 1977. Fort Collins *Coloradoan,* April 4, 1977; June 12, 1977.
99. *Triangle Review,* October 26, 1977.
100. Fort Collins *Coloradoan,* December 16, 1977.
101. *Ibid.,* November 10, 1976; June 23, 26, 27, 1977; March 21, 28, 1978. Denver *Post,* May 25, 1977.
102. Fort Collins *Coloradoan,* May 24, 1978.
103. *Ibid.,* February 14, 1978.

Chapter V

1. *Triangle Review* (Fort Collins), August 2, 1978.
2. Editor's Introduction, *American Behavioral Scientist,* 13 (January-February, 1970), 326.
3. Denver *Post,* August 3, 1976. Fort Collins *Coloradoan,* November 12, 1976; March 23, 1977.
4. Fort Collins *Coloradoan,* September 2, 1976. *Triangle Review,* September 8, 1976.
5. Thomas Ten Eyck, Oral History Interview (Oral History of Colorado Project, Colorado State Historical Society), April 24, 1975, pp. 7-8.
6. Fort Collins *Coloradoan,* September 2, 1976. *Triangle Review,* September 8, 1976.
7. Denver *Post,* July 30, 1978.
8. Fort Collins *Coloradoan,* March 31, 1977.
9. Denver *Post,* August 4, 1976. Loveland *Reporter-Herald,* August 14-15, 1976.

Name and Place Index

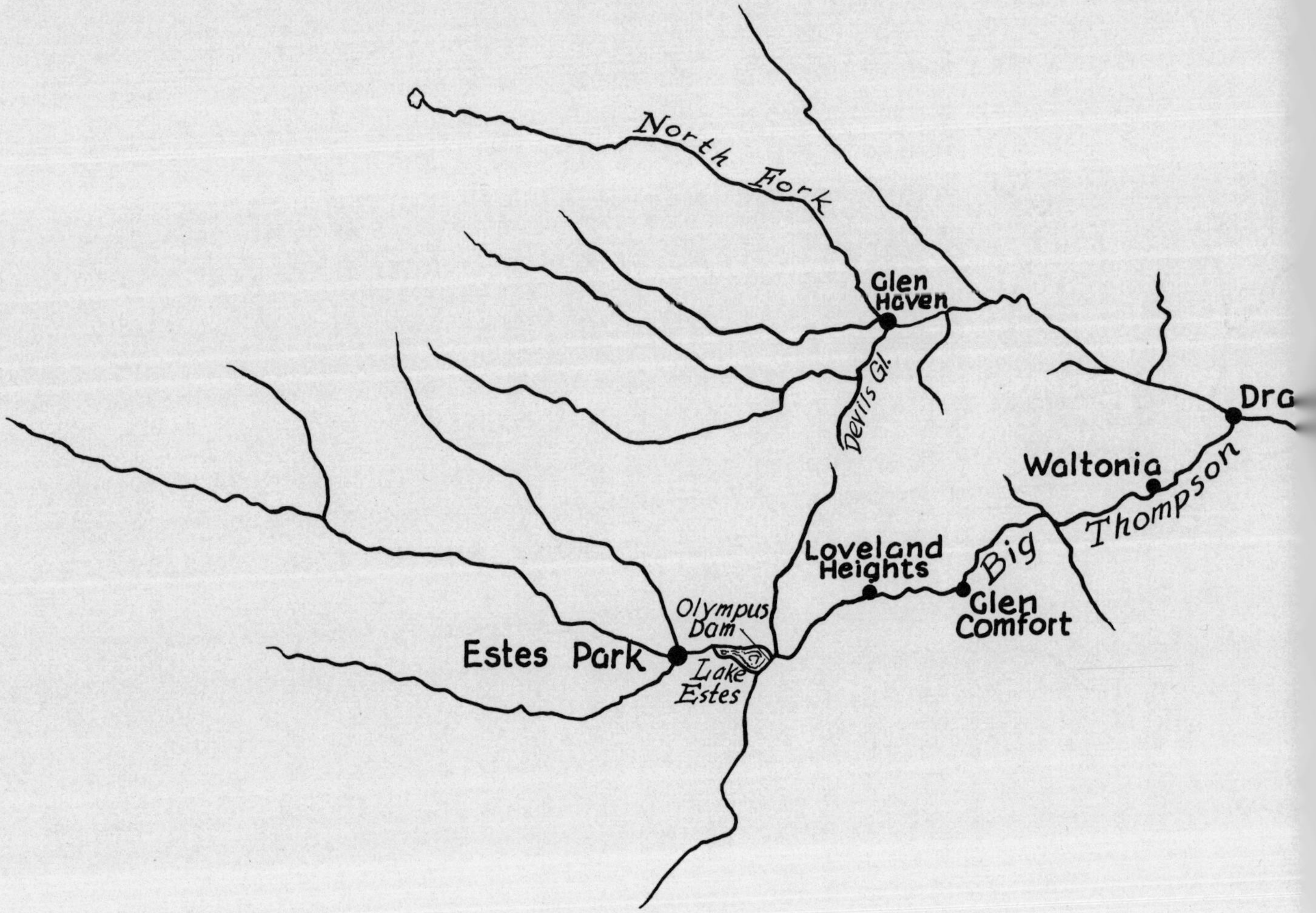
North Fork
Glen Haven
Devils Gl.
Dra
Waltonia
Thompson
Big
Loveland Heights
Glen Comfort
Olympus Dam
Estes Park
Lake Estes